We Discover

We Discover

Edited by Marc Guttman

Published by Marc Guttman
PO Box 623
East Lyme, CT 06333

This book is printed on acid-free paper.

Cover Design by Steven Mercado

ISBN 978-0-9849802-3-9

First Printing, 2016

This book is dedicated to Tara and Liam.
May they continue to question and to enjoy and endeavor in discovery.

Contents

Foreword: So Many Interesting Paths to Discovery

Marc Guttman

Marc Guttman works as an emergency physician and is the editor of two other books, Why Peace *and* Why Liberty. *He currently lives in Connecticut with his wife and children, and most enjoys spending his time with them and playing outdoors.*

There is an incredible photograph of paleontologist Walter Granger unearthing a Diplodocus lower extremity in Como Bluff, Wyoming in 1898. Not until I saw this picture as a young adult did I really begin to conceptualize the scenario of humans finding, for the first time, giant prehistoric fossils in the earth in the midst of the Industrial Revolution. I can only imagine how these discoveries created a perspective shift for so many as they helped inform our understanding of the planet. It must have been similar to people seeing the first images of Earth from space — a sight I've taken for granted since childhood.

The idea of human endeavor as the universe exploring itself, I have thought a fun perspective. As we hairy creatures on a wet rock hurtle through a galaxy and look down through microscopes, dig deep into the earth's crust, combine compounds in labs, implant bionics onto the human body, and shoot spacecrafts to outer space, the forces of the universe interact with each of us, as we influence it. It reminds me a bit of a boat chugging across an ocean, both affected by the sea's motions, currents, waves, and tides, while also contributing it's rippling energy to the whole.

When I was a student at university, I thought of contributing to the scientific discovery process. But, I had too many doubts. Going through the rigors and structure of school and study throughout my childhood, my mind was a good bit fastened, unfortunately. I was unsure whether I was capable of offering originality. My time assisting in a biochemistry laboratory didn't incite my interest much further. It wasn't until I began traveling independently, a time of endless leisure and adventure, that I started generating authentic ideas. I gained some perspective, a renewed interest in learning, and some self-confidence in my brain's ability. I was interested in work that was human-based, science-related, that I could employ around the world, and decided to start medical training.

Throughout my study I was fascinated by the discoveries in human physiology, from the many endocrine feedback systems that keep our chemistries in balance, our complex neurologic systems that allow us to experience our environment and interact with it, the hematologic system that supplies oxygen to our tissues, clots off bleeding sites, and provides for our immune systems that protect us from pathogens all about us, our reproductive systems that transport (and combine) the genetic material that carries all the coded information for the many different types of cells in our bodies to manufacture the necessary proteins and structures to perform all the functions for life, the psychology of the human mind, the environmental molecules we need ingest for survival and good health, to the symbiotic relationships we have with microbes. I learned the importance of understanding the ways these systems become diseased, and also interventions by which they can be restored and harmonized.

Medical science and technology is fascinating, and I enjoy learning medical history. The medical instrument I use most is the simple stethoscope. I use it about 20 times during a shift in the emergency department. I have auscultated the heart and lung sounds of more than 50,000 patients and counting. I can often gain insight into a patient's health by listening to his or her breath sounds, heart beat, and heart valves. Rene Theophile Hyacinthe Laënnec invented the first stethoscope in 1816 in Paris, where he used auscultation to diagnose a widespread illness of the time: tuberculosis, known then as consumption. He went on to document descriptions of various sounds from the chests of healthy and sick individuals. Perhaps not surprisingly, many physicians of the time opposed the use of the stethoscope for various reasons. They were apprehensive about the new technology and how to understand various thorax sounds. They were also uneasy about leaving behind traditional practices and incorporating new techniques in disease diagnosis, concerned about becoming too reliant on technology rather than history taking and clinical skills to diagnose pathology, and they were afraid of becoming (or being perceived) as a technician rather than a scholar and physician. But as more patients sought practitioners proficient in auscultation, the stethoscope became universally utilized in medical practice.[1]

In 1895, Wilhelm Conrad Röntgen had been experimenting with cathode rays. These are produced when electric currents were discharged in high-vacuum tubes—and he noticed that when they escaped would light up a fluorescent screen. He soon was creating "shadow pictures" with various objects. After seeing the outline of his own thumb and finger in a picture, he then asked his wife, Anna Bertha Ludwig, to place her hand on a photographic plate for fifteen minutes, taking the first X-ray of a human being. Otto Glasser, Rontgen's biographer, wrote "When he showed the picture to her, she could hardly believe the bony hand was her own and shuddered at the thought that she was seeing her own skeleton. To Frau Roentgen, as to many others later, this experience gave a vague premonition of death."[1]

Today medical practitioners can use several modalities to see inside the bodies of our patients for diagnostic and therapeutic purposes. I have a vivid memory of a 3:00 am procedure during a 36 hour shift as a medical student in the radiology department of an old New York City hospital. An interventional radiologist was placing a Greenfield filter in the inferior vena cava of a patient who was suffering with lower extremity veinous blood clots, but anticoagulation (blood-thinning) therapy was contraindicated, because the patient had bleeding risks. The filter was to prevent the blood clots from traveling up to the right-sided chambers of the patient's heart and then on to pulmonary arteries where it would cause respiratory and then cardiac failure.

I remember being exhausted, but feeling enthralled and privileged.

Today, one of my favorite modalities is using an ultrasound machine at patients' bedsides. I can quickly learn if a gallstone is obstructing the outlet of a gallbladder, if a fetus is appropriately implanted in the uterus, if a retina has detached, if a patient in shock requires more fluid resuscitation or rather vasoconstrictive medication by viewing the inferior vena cava, and where and how much a trauma patient may be bleeding around his or her heart or intra-abdominally.

Video: Inferior Vena Cava Filter Placement - http://tinyurl.com/ivcfilter

Pediatrician J.L. Wilson said, "Of all the experiences the physician must undergo, none can be more distressing than to watch respiratory paralysis in a child with poliomyelitis - to watch him become more and more dyspneic [air hungry], using with increasing vigor every available muscle of neck, shoulder and chin - silent, wasting no breath for speech, wide-eyed and frightened, conscious almost to the last breath."

Philip Drinker and Louis Agassiz Shaw developed the Drinker-Shaw respirator in 1927. The iron lung was an airtight container that enveloped the body below the neck. Pumping air from the container lowered the pressure within, causing the rib cage to expand and for air to flow into the lungs. When returning the internal pressure to atmospheric pressure, air flowed from the lungs. In 1929, Drs. Philip Drinker and Charles F. McKhann published an article entitled, *The Use of a New Apparatus for the Prolonged Administration of Artificial Respiration: I. A Fatal Case of Poliomyelitis*, that reported successful clinical testing of the Drinker-Shaw respirator. In the 1930s the machines were expensive, but the National Foundation for Infantile Paralysis raised hundreds of millions of dollars to support research, lend respiratory equipment, and create care centers for chronically disabled individuals. In the 1950s, intratracheal intermittent positive pressure ventilation improved polio patient care. In 1952 in Copenhagen during a polio epidemic, Bledgam Hospital had 900 polio patients, 75 of whom needed ventilation. This required 250 medical students to provide continuous manual ventilation. All medical school classes were suspended until the end of the epidemic. Ventilators that delivered automatic intermittent positive pressure were rapidly developed. [1]

Much of the above discussion on historical medical technologies was informed and inspired by Stanley Joel Reiser's terrific book, *Technological Medicine: The Changing World of Doctors and Patients.*

These technologies to diagnose and treat human (and animal) pathophysiology are extraordinary in their concepts and benefits.

I have enjoyed practicing medicine and applying the evidence-based knowledge gained by other physicians to best care for my patients, but I have never returned to the idea of contributing to scientific discovery. And, yet, I am fascinated and impressed by those who do, how they improve our understanding and wonder, and help us live more prosperously, healthily, and harmoniously. It is discovery and the wealth it creates that better allows us to nourish ourselves with healthy foods, have access to materials and energy, homes, and hot outdoor showers, educate ourselves and our children, provide for our health care and safety, protect our environment, aid others, and engage in the activities in which we delight.

I have long marveled that when scientists, innovators, and explorers discover and create it generally occurs on the margins. That is discovery, and all evolutionary processes for that matter, including those experienced in industry and in the social consciousness, always is achieved by those working off the knowledge and technology advances by others. And, it is most often incremental. Only infrequently are there moments of dramatic change and advancement. Even then, these times of rapid flux and progress occur after steady cumulative changes that have eventually led to the many toggles falling into their places. This all occurs across time and around the planet through both direct collaboration and indirect cooperation, those individuals and organizations throughout our global community trading freely on ideas, products, and services for their mutual benefit. And, it's beautiful.

Take for example Willem Kolff's original artificial kidney, 1943. Stanley Joel Reiser informs us, "Kolff was not the first person to attempt kidney dialysis. He neither introduced nor refined some of its essential elements, such as the use of cellophane membrane or its anticlotting agent heparin. But he made the conceptual, technical, and social connections needed to establish the innovation's usefulness." Innovators also must suffer all sorts of obstacles in their quests. As Reiser explains, "Kolff's work had to overcome not only the hurdles of a journey into the scientific unknown but also the trials of a wartime environment, which created shortages and tested courage. Shortages of rubber and glass tubing, essential in connecting patient and machine, and leaks in the cellophane tubing (once caused by dropping scissors on the kidney machine) were typical of the technical dilemmas Kolff faced. Equally difficult were Nazi threats and intimidation, which Kolff resisted. He went underground for a short time after refusing to sign a declaration of allegiance to Germany, and he was instrumental in keeping from Nazi hands a portion of the more than ten thousand men rounded up in Rotterdam by the Nazis to work in Germany. Kolff displayed courage in confronting and persuading the commander of this operation to allow him to choose and take charge of eight hundred men in most serious need of medical attention."[1]

In 1954, John F. Enders, Thomas H. Weller, and Frederick C. Robbins won the Nobel Prize in Physiology/Medicine for their cultivation of the poliomyelitis virus in tissue cultures back in 1948. They used penicillin, discovered by Alexander Fleming in 1928 and later developed by others, to prevent bacterial growth that otherwise would contaminate the tissue. This innovation enhanced Jonas Salk's research on the polio vaccine. He used formaldehyde to kill the polio virus growing in tissue, while keeping it intact to trigger a patient's immune response. He inoculated volunteers, including his wife, children, and himself. All produced antibodies without becoming ill. In 1953 Salk reported his findings: *Salk, J., Bennett, B. L., Lewis, L. J., Ward, E. N., and Youngner, J. S. --- Studies in human subjects on active immunization against poliomyelitis. 1. A preliminary report of experiments in progress. JAMA 151(13): 1081-98, 1953.* It was the start of valuable polio vaccination research.

I am appreciative of the tenacity of so many individuals. Professor Graeme Clark inventor and pioneer of the cochlear implant said, "I had much criticism and was referred to as 'that clown Clark,' but I was determined to persist and see it through, and I'm so pleased I did."

A friend of mine one night picked up a man hitchhiking back to our town in Colorado. The man was deaf and could read lips. My friend's car's inside light only went on when the driver's door was open, so in order for the man to understand him, my friend had to intermittently open his door while driving up a curvy, dark mountain road with an exposed cliff down to a river on one side. It's a comical story, but the hitchhiker, a local shop owner, would in a few years receive a cochlear implant and enjoy the sounds around us and verbal communication like so many others today.

Video: Cochlear Implants: People hearing for the first time! - http://tinyurl.com/cochlearfirst

I have imagined there is great self-satisfaction and joy in discovery and contributing to human progress, so I reached out to those engaged in it. And, I was right. This volume is an international collaboration by researchers, innovators, and explorers discussing the personal stories of the development of their interests, their research, discoveries, and creations, and their significance. They tell of their many diverse, often meandering, paths, as well as their trials, rejections, failures, and successes, their motivations and philosophies, moments of doubt and epiphany. They also share stories of when they experienced resistance from colleagues to their ideas and great collaborations. I have enjoyed learning from these individuals' personal accounts from around the world and about an extensive array of interesting fields of study, innovation, and exploration.

I am certain readers will enjoy the compelling journeys of *We Discover*'s contributors. You will travel to the prehistoric planet and the future, and all around the world. You will discover marine life deep in our oceans, insects in the grasslands, primates and predators in the jungles of Africa, the habits of whales and bears in the Arctic, glaciers in the planet's highest mountain ranges, and fossils in deserts and buried in the frozen mountains of Antarctica. You will be taken down into the microscopic world and up to the sun and way out to the outer solar system, and to a comet and a few planets in between. The neuroanatomy of the brains and emotions of humans and other animals are explored, as well as, the psychology of the human mind. Learn about the immortal jellyfish, what moss in the intestines of a 5,000-year-old iceman mummy discovered in the Alps informs us, the engineering of machines for astrophysical sciences and human biologic electromaterials, a good bit about human sexuality, and even real tractor beams!

And you will also learn about the human actors here; their motivations, obstacles, successes, failures, feelings, and perspectives. Find out what it is like to work in their labs and out in the field, on toboggans on the Antarctic snow or infusing carbon isotopes in a North American forest undergrowth. Experience who, what, how, where, and why we discover. These stories are narrative, informative, and philosophical, and all are entertaining. In the end, these are the goals of this endeavor: discovery, truth, wonder, progress, and enjoyment.

[1] Reiser, Stanley Joel. *Technological Medicine: The Changing World of Doctors and Patients.* New York: Cambridge University Press, 2009. Print.

Contact Marc Guttman at marc@wediscover.net

Tractor beams

David B. Ruffner

Originally from Manassas, Virginia, David Ruffner is currently Director of Analytical Services at Spheryx, Inc. He studied physics at Rutgers University, and graduated in 2009 with highest honors. He attended New York University for his graduate studies in physics, where he worked with Prof. David Grier, and studied optical forces in complex beams of light. David studied the role of polarization in optical force fields and how light can exert pulling forces in optical tractor beams. David was awarded a Dean's Dissertation fellowship and graduated in January 2015. At Spheryx Inc., David is leveraging his extensive experience with optical fields and optical scattering to commercialize holographic microscopy.

As a physics graduate student at New York University, I did research on tractor beams. If you aren't a science fiction fan, a tractor beam is a fascinating beam of light that can pull an object toward its source. Working with my advisor Prof. David Grier, we actually created a working, although microscopic, tractor beam. We demonstrated this beam experimentally by pulling little microbeads about the size of a bacteria over a distance about the width of a human hair. The New Scientist and other science news sources carried stories about our work.

Although it was amazing to be recognized for our work, the news articles didn't tell the story behind the story. I believe that knowing how discoveries are made is just as important as knowing what has been discovered. Discoveries don't come out of thin air (at least ours didn't!), and brilliant flashes of inspiration are not the typical mode of research progress. But knowing the real story of how a discovery was made can give insight into how to make a further discovery. So here is my story of how we figured out how to get light to pull things; how to make a tractor beam.

David B. Ruffner

It starts when I was a kid reading popular science stories. My godparents gave me a subscription to Scientific American magazine. Every single month I would read the magazine cover to cover. I loved learning about the big discoveries that were going to change everything. I was in for a bit of a shock when I started studying physics in college.

I had gone into college thinking that maybe I would be able to contribute a theory that would give a deeper fundamental understanding of the universe. I had read in Scientific American a number of articles about attempts to reconcile gravity with quantum mechanics. It seemed like we were close, so I figured that was something that I could do. Wow, the theory is much more complicated than I had thought! I did well in my classes so it may still have been possible to go into that field, but I started to realize that there are some very interesting physics at scales a lot closer to home.

With this in mind, I decided to study graduate physics at New York Univer- sity's Center for Soft Matter Research. Soft matter research focuses on material made from building blocks about the size of a biological cell. There is a lot of interesting physics because the building blocks are big enough to have complex properties but small enough to move around by thermal motion. Life also hap- pens at these scales (for the same reasons) so soft matter physics has strong connections with biology as well. I thought I would be able to do interesting research on topics that have more immediate application than quantum gravity.

This desire to study something more applied, concrete, and down to earth ironically led to studying tractor beams. At NYU, I especially liked Prof. David Grier's work. He is a world leader in developing laser tweezer technology. A laser tweezer can manipulate microscopic particles about the size of a biological cell only using light. It works by strongly focusing a powerful laser beam with a microscope. The focused spot of very intense light attracts biological cells or glass micro-beads and can hold them in three dimensions. I found it fascinating that light can exert forces and potentially be very useful. So in spring of 2010, I joined the Grier group, and started studying how the polarization of the light can be harnessed to create new forces in optical tweezers.

David B. Ruffner

In November of 2011, NASA announced that it was exploring practical implementations of tractor beams. Their initial search was based on progress made by three research groups including David Grier's. David Grier and Sang-Hyuk Lee had demonstrated pulling action over 8um using a spiral shaped or solenoid beam [4]. Jun Chen had theoretically shown that beams of light called Bessel beams could pull particles [1]. Finally Shvedov and his group had shown that they could control microparticles over a meter by heating them unevenly with a doughnut shaped beam of light [8].

I couldn't believe that NASA was interested in tractor-beams. All of the results were for relatively short ranged beams. The fact that NASA gave tractor- beams serious consideration gave me new appreciation for how blue-sky NASA research can be. I was skeptical that it really would be possible to make a long range tractor beam. But I was excited also because, despite the difficulties, there may be a way to make practical tractor beams.

The team from NASA contacted David G. to see if he would do further re- search on tractor-beams. He was interested, and I was the natural person in the Grier group to study tractor beams since my research had already been focusing on optical forces. David suggested that I work on making a simpler more robust version of the solenoid tractor beam. Despite my skepticism, I thought that David's idea was relatively straightforward to try and might even give me a chance to help make the discussion around tractor beams more grounded. I decided to give it a shot.

Let me first describe how we actually made these complex beams of light with our holographic optical trapping setup. First we used a powerful green laser as the source of our beam. Our laser is 5 watts, which is about a 1000 times more powerful than the typical laser pointer (we use laser safety goggles!). The light from this laser is directed with mirrors to a special device called a Spatial Light Modulator (SLM). This device is essentially a screen that can imprint a picture, or hologram, on the beam of light. Finally the light is tightly focused by a microscope objective onto a sample on a microscope slide. Depending on the hologram imparted by the

SLM, the light will focus to a single point, a variety of points, or even focus into rings or spirals.

The amazing part is that these beams are so intense that they can move and exert forces on little particles. But how can light move something? I think that is very counterintuitive and very cool. Light carries energy (for instance the warmth you feel from sunlight), but it also carries momentum. One clear example of light's momentum is how comet tails always point away from the sun. The photons of light hitting the dust of the comet tail exert a real measurable pressure mass are related. A consequence is that a wave carrying energy has to carry at least a little momentum as well.

In the lab, we can directly observe optical forces. We would prepare a sample containing little microscopic glass or plastic beads floating in water sealed between a microscope slide and a coverslip. The beads were one millionth of a meter in diameter, just big enough that we could see them in our research grade optical microscope. We would then shine the laser beam through the microscope and focus it in the sample. If the focal spot of the beam came near a floating bead then the bead would be pulled into the center of the beam and held there by the light. Alternatively, we could use the SLM to make rings of light that drive the beads around and around making a microscopic mixer.

One of the more interesting beams of light we make are the optical solenoid beams that I mentioned earlier that can act as tractor beams. Instead of focusing at a single focal spot, this beam focuses into a spiral that winds around the optical axis. When we focus one of these beams on a sample, a microbead will be attracted to the spiral. Once captured on the spiral the microbead can slide along it. The amazing breakthrough of Lee and Grier was showing that it's possible to direct the light in such a way that the particle gets pulled up the spiral against the flow of the beam [4]. It sounds impossible, but it works in a similar way to how a sailboat can travel against the wind by tacking. The curl of the solenoid allows the microbead to effectively tack against the flow of the light and be pulled upstream.

We wanted to see if we could extend these beams to longer range than the 8um that Lee and Grier had demonstrated [4], but we were held back by the complexity of these beams. To create these spiraling beams, we actually needed to added up a whole series of more fundamental beams. These building-block beams, called Bessel beams, look like a rod of light surrounded by a series of rings, and they have the interesting property of not diffracting over a limited range. The complexity of this series of beams made it very difficult to extend the range. Instead we tried to make a minimal version of the beam made from only two Bessel beams.

After working out the math, I was able to make these minimal solenoid beams, but I was unable to make them pull particles. I tried all sorts of parameters in the model, and had many conversations with David. However, when I projected these simpler versions of spiral light beams, they always pushed the micro-beads.

I felt discouraged, and I was unsure whether it was better to give up, or to stay optimistic and keep trying. The pulling force of the minimal solenoid beams was proving to be very weak if it existed at all. It didn't make sense to me to keep trying the same thing, but I didn't want to give up either.

Out of desperation I adjusted the parameters until the beam was not a spiral at all but a series of bright spots in a row surrounded by rings of light. The particles get attracted to the bright spots instead of getting pulled along the length of the beam. I realized that once the particles were trapped in the bright spots, I could shift the bright spots and move the particles with them. The beam acts like an optical conveyor belt, and can either push or pull particles. I tried it out and it worked!

I was cautiously excited that I could pull particles in this way, but it didn't seem very different than regular laser tweezers. A laser tweezer can also trap a particle in its bright focal spot. It can then pull the particle if its focus spot is shifted along the optical axis. Luckily, David saw a lot of potential in these optical conveyor beams, and even recommended that we apply for a patent before we write a paper about them. I also was putting together a proposal to my thesis committee about what my thesis would be about, so we threw in some figures showing how these new types of beams work.

As often happens in science, I came across another paper that seemed to have already done the exact same thing. In a series of papers

Cizmar had moved micro-beads using precisely the same type of beam that we used [2,3,9]. Sometimes it seems like it's impossible to come up with anything new, since there are so many smart people doing science. I thought there was no point to continue trying to do the same thing that was already done, and actually stopped working on the project for a while. We were working on another paper related to forces from the polarization of the light so I was busy anyway.

As we were wrapping up the other paper, I decided to revisit the optical conveyor beams. Upon closer examination of the paper by Cizmar and Zemanak, I found that they hadn't quite been able to pull the microbeads with their conveyor beam [2]. They had to use a flow of water to balance the radiation pressure of the light. Our beams were able to pull particles without any helping forces. The difference between our beam and theirs was that ours was more strongly focused. So we did have something new to contribute!

This observation gave me renewed energy to finish the paper. I worked hard on the measurements and found a set of parameters for the beam that worked really well. We were able to pull particles over a range of 33um. Though still microscopic this range is 4 times further than the previous record holder, the optical solenoid beams. Writing up these measurements and others, we published a paper about the optical conveyors as a new class of optical tractor beams [6].

Immediately after the paper was published, it set off a wave of articles in popular scientific magazines. The New Scientist Magazine published the first article a day before the paper was officially published. Next came Phys.org, the Huffington Post, Ars Technica, the Register, and others. We were even interviewed by Michio Kaku for his radio show Science Fantastic! It's amazing how eager people are to hear about real life tractor-beams even if they are only microscopic.

I was happy to receive recognition for this work and that people were so interested, but I felt like people were getting a somewhat distorted picture. First of all these short news reports didn't do justice to the papers that came before ours that set the stage for our discovery. It's understandable that they don't have the space to be able to do this, but it made me feel a bit embarrassed. Secondly, the main interest for regular people was whether these tractor beams could be scaled up to pull spaceships and things like that. But there is a gigantic difference between moving the microbeads over microscopic distances and moving spaceships over kilometers. I didn't know if our optical conveyor beams could be scaled up, or even if they were actually better than the optical tweezers. I wanted to put our results on a firmer footing.

So we set out to write a follow up paper about the optical conveyor beams to flesh out how they work as tractor beams. We planned on comparing the trapping properties of these optical conveyors to the theoretical predictions. Then using the understanding from this theory, we hoped to optimize the beam to greatly extend then range.

Things didn't turn out like we planned. We did find that the optical con- veyor beams were better at trapping the microbeads than the optical tweezers. However, the measured results did not agree with the theory we were using. They weren't even qualitatively right. And furthermore, I found that the previous results were pretty much optimal as far as the range of the optical conveyor beam. The only way I could increase the range was by using a larger microscope objective, which was able to pull micro beads over a range of 66um. These results were interesting, but we needed a correct theory in order to make a coherent story that we could write up. I was frustrated, because I was hoping to graduate soon, but felt like I needed to finish this paper before I could move on.

So we kept on pushing forward with the optical conveyor beam research. I would take breaks to work on my other project about the polarization of the light, but I would always come back to the optical conveyors.

Spring of 2014, I tried a new set of calculations to try to understand the optical conveyors. We had been using an approximation that made the math tractable, but that didn't match the experiment. My advisor had suggested against using the more complete theory because it's complexity makes it difficult to extract any simple relationships. I felt that it was important though, and got the OK from my advisor to work on it. I spent over a month coding up programs to perform calculations using the complete theory for optical forces on the microspheres. After all of the debugging and testing, I compared this more complicated theory against the experiment. It matched up!

With those theoretical results, we had a story to write up. We were able to show theoretically and experimentally that the optical conveyors beams have superior trapping properties than the optical tweezers. The excellent trapping properties depend on the interference structure of beam. The optical conveyor beams work especially well for particles smaller than the wavelength of light. Furthermore, our theory for the trapping properties enabled us to make predictions for the ultimate range of the optical conveyor beams. For small particles the beams can work even at low convergence angle, or long range.

I felt that this discovery gave tractor beams a much firmer basis; it convinced me that tractor beams provide real advantages over conventional laser tweezers. In our previous paper we could only suggest why optical conveyors were better than optical tweezers. Now, we were able to demonstrate quantitatively the advantage of optical conveyors over optical tweezers and explain why they work better. This new knowledge provided answers to many of the questions people had about tractor beams, like how far can they pull things and how big of a thing can they pull.

However, when we tried to publish our results we were rejected. The journal argued that our results were not sufficiently different from our previous paper. It's funny because I felt like this follow up paper on the optical conveyors had much more substance than our initial paper. If you were actually trying to design a tractor-beam, our second paper about optical conveyor beams would be much more useful.

We edited the manuscript and submitted it to a second journal, and this time we were accepted [7]. There was not press or interviews about this paper, but I was happy to have published it. It meant I was closer to graduation, and the paper would make for a really nice chapter in my thesis. I spent the fall writing up my dissertation [5], and then successfully defended in January 2015.

From my experience with the optical conveyors beams, breakthrough discoveries are more a comprehension on the part of the public, rather than a huge discovery on the part of the researcher. It is the small incremental advancements that keep progress going. The optical conveyors received recognition, but they were not the first tractor beams. Laser tweezers could already pull particles even if they are not technically considered a tractor beam. Lee and Grier had already tested the optical solenoid beams, which are definitely tractor-beams despite their relatively short range. Using the optical conveyor beams as tractor beams was a small step forward from the optical solenoid beams. However, it crossed a threshold that made the public realize that tractor beams are not just science fiction. This realization by the public made it a big discovery.

I learned from my advisor that one of the keys to making discoveries is to be relentlessly optimistic. I need to see the potential and the good in what I discover. With persistence in the lab, the improvements and small discoveries come even if they come irregularly and unexpectedly. But most of the time these little discoveries are only new to me. However it pays to be optimistic that the thing I discovered may actually be new to the world too. I run the risk of looking stupid when people point out that this has been known for hundreds of years. But I've found that often there is a little bit that's new even if it's very similar to what was done before. Sometimes that little bit crosses a threshold, making that little discovery into a breakthrough.

References

1. Ashkin, A., Dziedzic, J. M., Bjorkholm, J. E., and Chu, S. Observation of a single-beam gradient force optical trap for dielectric particles. *Optics Letters* 11, 5 (May 1986), 288–290.
2. Chen, J., Ng, J., Lin, Z., and Chan, C. T. Optical pulling force. *Nature Photonics* 5, 9 (July 2011), 531–534.
3. Čižmár, Tomás and Garcés-Chávez, Veneranda and Dholakia, Kishan and Zemánek, Pavel. Optical conveyor belt for delivery of submicron objects. *Applied Physics Letters* 86, 17 (2005).
4. Čižmár, Tomás and Kollárová, Věra and Bouchal, Zdeněk and Zemánek, Pavel. Sub-micron particle organization by self-imaging of non-diffracting beams. *New Journal of Physics* 8, 3 (Mar. 2006), 43–43.

5. Čižmár, Tomás and Šiler, M. and Zemánek, Pavel. An optical nanotrap array movable over a milimetre range. *Applied Physics B* 84, 1-2 (2006), 197–203.
6. Grier, D. G. A revolution in optical manipulation. *Nature* 424, 6950 (2003), 810–816.
7. Lee, S.-H., Roichman, Y., and Grier, D. G. Optical solenoid beams. *Opt. Express* 18 (2010), 6988–6993.
8. Ruffner, D. B. Optical Forces in Complex Beams of Light. PhD thesis, New York University, 2015.
9. Ruffner, D. B., and Grier, D. G. Optical conveyors: A class of active tractor beams. *Phys. Rev. Lett.* 109 (2012), 163903.
10. Ruffner, D. B., and Grier, D. G. Universal, strong and long-ranged trapping by optical conveyors. *Opt. Express* 22, 22 (Nov 2014), 26834–26843.
11. Shvedov, V. G., Rode, A. V., Izdebskaya, Y. V., Desyatnikov, A. S., Krolikowski, W., and Kivshar, Y. S. Giant optical manipulation. *Physical Review Letters* 105, 11 (2010), 118103.

Unseen Connections

Suzanne W. Simard

Suzanne W. Simard, professor of forest ecology in the Department of Forest and Conservation Sciences at the University of British Columbia in Vancouver, Canada, is known for her research on mycorrhizal networks, which are characterized by underground webs of fungi that facilitate communication and interaction between trees and plants of an ecosystem. Other areas of research include forest stand dynamics, ecological resilience and global change.

The woods are lovely, dark and deep. But I have promises to keep, and miles to go before I sleep. - Robert Frost

My kid-brother, Kelly, runs down the gangplank onto the logger's houseboat yelling "Jiggs fell in the outhouse!" We can hear Jiggs' muffled howls coming from the earthy pit. The howls are mostly in fright but also tinged with embarrassment, knowing his canine curiosity has finally landed him in the deepest possible shit. "Tabernac!" Uncle Wilfred growls, as he strides up the path to the outhouse with a shovel. By this time, my whole family -- Mum, Dad, sister Robyn, Kelly and I -- are peering down through the hole in the seat, calling to Jiggs that we will surely rescue him. Jiggs' beagle eyes, partly covered by toilet paper and drooping with remorse, stare back up at us. Uncle Jack, with half his fingers missing from chainsaws and axes, joins the rescue operation with a heavy mattock. He is laughing, as always. Rich smells drift up and we break into peels of laughter. The men quickly organize for the messy excavation. Soon remains just the men and I, them digging and me watching because I have a particular fascination with earthy detritus. I become entranced as the men dig down through the layers, starting with the dark forest floor, then the white then red then yellow mineral soil horizons underneath. The men curse. The fine roots have formed an impenetrable mat and the coarse ones are jutting at awkward angles deep in the soil. These roots provide Jiggs a foothold to rest from his struggles. The interwoven pallet of roots and soil also serve to anchor and connect the colorful mix of birch, hemlock and cedar trees. This diverse mixture inhabits the inland rainforests of British Columbia's Monashee Mountains. Then, with a great deal of howling and cursing en français, the men liberate Jiggs from his earthy bind. Thus began my serendipitous journey of discovery about the unseen world of soil creatures connecting trees in the forest.

The forest seems part of my blood and bones. The emerging field of epigenetics has revealed that good and bad experiences in life can tag our DNA, and these changes can be copied down through generations. I am convinced this genetic process has embedded forests into my soul. My Dad would often take us up the mountain above the outhouse, following an old log flume deep into the forest. At the top of the flume, the ephemeral springs in the cedar swales would greet us with sweet aromas of wild ginger and skunk cabbage. There, Dad told of his own father and uncles who, in the 1940's, hoisted old-growth cedar logs into the flume with their pike poles, chokers, and horses. They were hand-fallers, selecting only a few to eke out a living. My Dad was a kid in those days, and his job was to stand on the log boom in Mabel Lake at the base of the flume. He waited for the logs to come barreling down the mountain, and when they pierced the water, they would disappear deep into the lake leaving an eerie silence. Where would they surface? Would they impale him from below? What took only seconds seemed like hours of torture. When they did emerge, it was as though a gun had been fired and the enormous logs shot straight into the sky. Boom! Splash! My frightened Dad would then use his pike pole to train the bobbing cedar logs into the growing boom. The fear of this experience in my dad, along with the wonderful aroma of wild ginger root, are etched in his DNA. These stories and experiences of the forest must have then been copied into my DNA.

These early childhood experiences with my logging family in the forest-clad mountains of interior British Columbia eventually drew me to study forestry at the University of British Columbia. After graduating in 1983, I worked as a silviculturist in the Lillooet Ranges for a logging company, Evans Forest Products. There, I sometimes worked alongside

Grant Hadwin (http://tinyurl.com/goldenbough), forest road engineer, who was the eventual fugitive hewer of the famous Golden Spruce on Haida G'waii. He acted in protest against the powerful people in control of the provincial harvest. Like Grant, I was horrified by the relentless clearcutting, and I was deeply conflicted by my part in it. Back home, I watched with heartbreak as my beloved Monashee forests were carved into uniform square clearcuts, which eventually became so numerous that the old-growth forest was all but gone. These clearcuts were replanted with fast-growing lodgepole pine and Douglas-fir instead of the richly diverse, multi-successional native mixtures. Silviculture policy in the mid-1980's, in place to this very day, encouraged removal of my beloved birch, hemlock and cedars to eliminate their competitive effects. These species were cut, hacked and sprayed to promote fast growth of the commercially valuable pine and fir, which were slated for short rotation harvest. The growing wisdom and scientific evidence that such simplified forests were less productive and resilient than more diverse forests was discounted. Ignored were the increasing infestations of insects and diseases in the newly cleansed plantations. Out of frustration, I returned to university to earn graduate degrees in forest ecology, the same year Bruce Cockburn released his song, "When a tree falls in the forest, does anybody hear?"

Through my graduate research, I learned that most of the forest is unseen. What we do see -- the massive tree trunks and myriad of arboreous creatures -- is simply the tip of the iceberg. Underneath is an 'Otherworld', where creatures of life and death drive biogeochemical cycles. I joined scientists who were probing the soil with lenses, microscopes and later DNA markers and learned that the unseen Otherworld is teaming with communities of bugs of all sizes. We now know soil as a complex social system, where creatures of the soil foodweb eat each other, communicate, exchange resources, transmit warning signals, and even eavesdrop on each other. This living system is energized by the sun. Solar energy is transformed by plants into photosynthate carbon, which is exuded from roots to the rhizosphere, which is then reduced in a cascade of biochemical reactions to forms usable by creatures of the dark. Some of the rhizosphere creatures form symbioses with plant roots, some live off the sloughed carbon detritus, and still others are free-loaders. One of the symbioses, mycorrhizas (literally fungus-roots), are mutualisms where plants trade photosynthate for scarce nutrients that the fungi acquire from the soil. These mycorrhizal fungi form tiny mycelial threads that penetrate even the smallest soil pores, breaking down soil organic matter, taking up nutrients and water, and physically connecting plants in vast underground networks.

For my doctoral research, I went straight to my roots. I wanted to understand why diverse forests – namely mixtures of trees – were more resilient and productive than monocultures. Why did plantations of Douglas-fir have less root disease when paper birch was allowed to mix in? I was driven by a vision of sustainable stewardship of my home forests. I was a government forest scientist at the time when Drs. Melanie Jones and Dan Durall, both mycorrhizal ecologists, contacted me about possible collaborative research on this topic. They explained that Sir David Read and colleagues in the UK had just published a paper in Nature showing that roots of lab-grown pine seedlings had become interconnected by mycorrhizal fungi, which served as a sort-of pipeline for photosynthetic carbon to pass from seedling to seedling. Amazingly, the more the subordinate pine seedling was in the shade, the more carbon it received from its neighbor through the mycorrhizal network. I thought, Eureka! This could help explain why Douglas-fir can thrive in the shade of paper birch. Testing this concept in real, live mixed forests became the central focus of my doctoral research.

This research involved isotopes. My colleagues and I labeled birch, fir and cedar seedlings with stable and radioactive isotope markers, which are rare forms of carbon that allowed us to trace photosynthate-derived carbon as it moved through the mycorrhizal fungal network. We used two isotope markers because we specifically wanted to detect if carbon was moving back-and-forth between tree species through the fungal links, and if one of the tree species had a net gain.

Accomplishing this objective deep in the wilderness was no small feat. The radioactive carbon had to be kept frozen in vials using liquid nitrogen, which involved driving back-and-forth daily from the bush to the closest university lab 300 kilometers away. These frozen vials were taped to the inside of gas-tight plastic bags encasing the seedlings growing in the forest.

To release the isotope as tagged carbon dioxide that the tree could then photosynthesize, we had to inject acid into the frozen radioactive base using a large glass syringe. Once the needle was plunged into the bag's septum, the acid was slowly dripped into the frozen vial. I was truly frightened by this procedure. What if I breathed in the radioisotope and poisoned my body forever? Would I die a slow death of cancer? Even thousands of years after death, would my bones still be glowing? To abate my fear, I wore a hazmat suit, giant plastic goggles, and rubber gloves and socks sealed with duct tape.

Wearing this cumbersome gear, I waddled from tree to tree, injecting acid, while Dan manned the cooler of frozen vials. I sweated profusely under the hot mid-summer sun. As I stumbled to the furthest set of trees, I suddenly heard snorting. I knew that sound. I looked up to see a mother bear and a cub walking toward me. I looked at them, then the syringe in my hand, and I thought – "OH NO!" I remembered – if its grizzly brown, hunker down; if it's black, fight back. In a frightful moment, I did neither. Instead, I ran, holding the syringe high in the air, shouting "GO AWAY!" The bears ran after me. I ran faster. They caught up to me. Just as the bears and I were approaching the work truck, Dan jumped onto the roof and I clamored up behind. Mama bear stood on her hind feet a few meters away and curiously stretched her neck to see what sort of beast we were standing on. We shouted. We waved our arms. I warned her I had radioisotopes that I could squirt at her if she wasn't careful. She shrugged and grunted in amusement, somehow seeing my bluff that I had only acid in the syringe. After several moments of standoff, she sank back to her all-fours, gestured to her cub, and they both slipped silently into the woods.

My colleagues and I eventually published this work in our own *Nature* paper, showing that Douglas-fir and paper birch were intimately interconnected in a diverse mycorrhizal fungal network. Even more, we reported that photosynthetic carbon moved back-and-forth between the two tree species through this network, but with net transfer from birch to fir. The net gain in carbon by fir was enough for the trees to make seeds and reproduce. We also found that the more shade birch cast on its fir neighbors, the more carbon it donated. The cedar received little isotope from the networking birch and fir because this species forms an entirely different class of mycorrhizas.

This was a major advance on Sir David Read's work because we had demonstrated a dynamic exchange of substantial amounts of carbon between two different species in the wild forest. Eventually, we discovered that the direction of net carbon transfer changed over the growing season, with Douglas-fir sending some of its carbon back to birch in the spring and fall when birch was leafless.

This dynamic exchange between tree species, shifting with environmental conditions, suggests the trees are attuned to each other. This sensitive dynamic among species is essential to maintain balance in the forest. Our findings challenged prevailing ecological theory, which said that forests are a dog-eat-dog world. In other words, that inter-tree competition reigns superior and cooperation is rare or inconsequential. With our discovery, I felt armed with a scientific basis not only for a shift in this theory but also forest policy.

I wasn't prepared for what happened next. I had returned from graduate school to my government position and naively requested a meeting with the forest policy makers in the provincial capital. I was buoyed by interviews with the media over our *Nature* paper and was undeterred by the sphinx-like silence from government. Surely I simply needed to explain the research, show them the evidence in the forest, and then help map a course for change to forest policy and practice. We convened for a half-day conference and field trip, to which dozens of policy makers, foresters and scientists traveled from around the province. I designed the field trip to show a range of plantation conditions – "the good, the bad and the ugly." My presentation was quickly met with resistance and some minor heckling from a young forester, but I remained optimistic and determined. The next day in the field, we stopped first at a dense young forest of paper birch and Douglas-fir to observe "the ugly." "This obviously shows birch kills conifers," the head policy-maker muttered. Nods from the young heckler and other policy makers revealed a consensus amongst them. Still, we pushed on and visited a wide range of plantations. We discussed the inner workings of the ecosystem, including the multiple interspecific interactions that shape the forest. By the time we got to "the good," the primary architect of the regeneration policy had run out of patience. Towering over me, he shouted that I simply did not understand the forest or forest policy. Planted forests must be weeded and simplified! I felt the other policy men close in around me. I didn't hear what was said next, but I stepped back from the crowd. The field trip was over. It triggered a letter of reprimand on my personal file and tight scrutiny of all of my subsequent research. I had been silenced. Within a couple of years, I left that position for a professorship at UBC. But most interestingly, the same summer I left government, the policy makers revised their regeneration policy, effectively reducing by half the amount of herbicide spraying in the province.

Academia has since provided me with immense freedom to set the course of my own research program. Using an alchemy of imagination, intuition and experience, I have made further discoveries of the unseen connections in forests. We have found in Douglas-fir forests that conspecific seedlings establish into the mycorrhizal network of old-growth trees, which provides them with sufficient resources to survive. These networks form a "scale-free" pattern, such that the large, old trees are the hubs of the network and the younger trees the satellite nodes, much like the pattern of a modern-day transportation network. We found that old trees rapidly transmitted carbon, nitrogen and water to the seedlings, increasing their nutrition, survival and growth. In drier climates where the forests experienced drought stress, old trees transmitted more water to connected seedlings than did trees in wetter forests where they were replete. Thus, this intricate below ground telecommunications system appeared essential for the recovery and resilience of the forest under stress.

We are continuing to learn that the old hub trees, or "mother trees," can distinguish kin from stranger seedlings. When they recognize neighbors as kin or strangers, the mother trees adjust their competitive and cooperative interactions to favor kin. In our experiments, mother trees shuttled more nutrients and supported greater mycorrhizal networks to close relatives, thus providing a competitive edge and greater survivorship of their kin. The mother trees also shared small amounts of resources with strangers, suggesting selection mechanisms may exist not just for species, but also communities. Moreover, in keeping with our earlier work showing increased mycorrhizal fungal network facilitation along climatic stress gradients, we also found greater kin cooperation with increasing drought.

Mycorrhizal networks have also served like telegraphs for transmission of biochemical signals. We have recently discovered that injury to one tree resulted in the transmission of defense signals through the connecting mycorrhizal mycelium to neighboring trees, even though they were a different species. These neighbors responded with increased defense-gene expression and defense-enzyme activity, resulting in increased pest resistance. We discovered this by defoliating a Douglas-fir tree with insects then measuring RNA gene expression and the enzyme concentrations in neighboring ponderosa pine trees that were connected belowground by a mycorrhizal network. Sudden injury to the Douglas-fir not only triggered a defense response in healthy pine neighbors but it elicited a rapid transfer of photosynthate carbon from the fir to pine. It appears that the dying trees not only transmit signals but also energy through fungal networks to neighbors for the benefit of the forest community. This phenomenon has evolutionary benefits not only for the trees, but also the fungus because it increases the security and diversity of carbon sources.

Many new questions arise from these findings. Could communication between trees play a role in reducing other kinds of stress, such as drought or root disease? Could protecting these communication pathways increase the resilience of forests to climate change? What are the signaling molecules involved in defense communication and kin recognition? Are defense signals, as with kin recognition, preferentially transmitted to kin to improve their survivability when a forest is stressed? Is epigenetics involved in hardwiring these responses in the DNA of future generations? If so, can we manage mother trees in regenerating forests so that future generations are more resilient to climate change?

As I was eagerly learning about the importance of connection and relationships to the health of forests in my research, I received the devastating news in 2012 that I had breast cancer. My mind went immediately to the radioisotopes. But no, it was likely caused instead by the immense stress I had suffered during the breakup of my marriage only a few months earlier. I was treated with a double mastectomy, chemotherapy, radiation and hormonal therapy during the early part of 2013. During these debilitating treatments, I was forced to completely rewire my thinking about my life and relationships. I reached out to strengthen bonds with my family and friends. Daily walks in my beloved forest were also essential to my recovery.

That fall, although still weak from chemo, I visited my favorite forest field experiment with my best friend, Jean Mather. Jean was my first teacher of the beauty of forests, through countless earlier days of backpacking and skiing together in the woods of British Columbia, earning her the nickname "Mather Nature." This post-chemo fieldwork with Jean was a healing time, and we spent a week together laughing and measuring trees. I had established this experiment 27 years earlier during my Masters research, at the same time the province had started its killing spree against all plants broadleaved. I sought to understand how Sitka alder, a native nitrogen-fixing shrub that is abundant in the vast montane forests of British Columbia, interacted with planted lodgepole pine. Because nitrogen is the most limiting soil nutrient in our forests, I was worried that the zealous removal of alder would undercut the natural ability of the forest to recover from clearcutting. Understanding these interactions involved some unusual measurements.

Jean and I laughed when we remembered one of those measurements requiring a midnight visit. On this occasion my dear father was my assistant. We drove into the deep dark forest and hiked an hour through the bush to the experiment, carrying our heavy instruments. One of the instruments was a pressure bomb, the other a cylinder of nitrogen gas, used together to measure the degree of drought stress the seedlings were suffering. Dad pretended to be brave but I knew he was just as jittery as me. As we approached the experiment, we sensed a large animal nearby. Snap! My heart leapt. I heard Dad suck in a sharp breath. Then, a cow bellowed "mooooo"! We both laughed nervously and trudged on. I set up a measurement station in one of the alder treatments and instructed Dad to cut off a lateral shoot of a planted pine and return it to me so I could take the water stress measurement with the pressure bomb. "Take the lateral, not the leader!" Without a leader shoot, the pines won't know how to grow tall," I explained. Dad nodded. He looked scared but really wanted to be a brave helpful soldier. He disappeared into the dark, then returned a few moments later holding a big juicy leader. Dad!?

In the early years of this experiment, removal of the native alder community actually improved the physiology and growth of the planted pine, even though more seedlings died and nitrogen availability in the soil was declining. The early loss of this soil nitrogen, however, foreshadowed what happened next. Using dendrochronology in 2013, we discovered that the growth of pine with no alder neighbors had collapsed during a previous three-year drought. We found these trees also suffered nutrient deficiencies, likely due to the loss of the nitrogen-fixing bacteria and mycorrhizal fungal symbionts normally hosted by the alder. A few years later, the mountain pine beetle outbreak, famous for damaging 47 million hectares of pine forest in North America over the past decade, reached my experiment. The pine trees where we had removed the alder were infested and killed. Incredibly, pines where we left the native alder community intact continued to flourish and none died.

The cumulative stresses of poor nutrition, climatic drought and insect infestations in stands where we had removed alder had crossed a tipping point. The stressed monoculture stands died; the vigorous mixed stands thrived. In the mixed stands, the alder had played a synergistic role in bolstering resistance of the pine to stress, similar to civil societies. The irony of this was not lost on me. Like trees, humans can only take so much stress before they succumb. This research invites new questions about the role microbial symbionts play in bolstering resilience of forests to climatic changes.

This story of scientific discovery about connectedness in forests is interdependent with my human story. My own deep early connection to the forest, and my personal and professional experiences that exposed the necessity of maintaining this connection, were foundational to the questions I was able to eventually ask as a scientist. My scientific discoveries have since unfolded like a treasure hunt, where one clue, one happenstance, and one experience led to a discovery which led to yet another clue, another discovery, and so on. It is clear to me that serendipity, not robotic hypothesis testing, was essential for revealing the unseen connections in forests. Who I am, what I experience, and what I discover has led to the next questions to explore. I am delighted that our discoveries about connection and communication in forests resonate with people. People seem cheered with this new knowledge. They laugh, smile, and sigh with relief. I think it makes them feel hopeful – knowing that nature has evolved with complex mechanisms for coping with stress. The research we have conducted is only scratching the surface of how nature works. But as we peek into the inner workings of forests, I have become convinced that nature will help us heal and adapt to climatic changes, provided we treat it with respect.

Having a Look: Discovering and Exploring the Outer Solar System

David Jewitt

David Jewitt is a professor of planetary astronomy at the University of California, Los Angeles. He focuses his attention on the smallest, least evolved solar system bodies because these are the most likely to preserve information about the origin of the planets. His work has been recognized by several national science academies and by major science prizes.

Discovery

George Orwell wrote that "To see what is in front of one's nose needs a constant struggle," a statement that is as true in the physical sciences as in Orwell's arena of social commentary. Scientific discoveries can take many forms, from the unveiling of particular objects, to new classes of object, to finding new phenomena or new physical laws. Some discoveries are expected, while others are made because new instrumentation enables a measurement that was not previously possible. But surprisingly often, discoveries occur because we break through a kind of "perception barrier;" something that was unseen in front of us all along is noticed for the first time.

When we're young, making discoveries comes naturally. That's because curiosity is the number one survival tool of our species. We're programmed by evolution to spend our first years filling our brains with countless, practical discoveries about the external world. Childhood is effectively a period of intense discovery that equips us to survive and thrive on a complicated and dangerous planet. These are the roots of science, which clearly stems from our need to understand the threats posed, and the resources and opportunities presented, by the natural world.

Sadly, the older we grow the more likely we are to settle into a comfortable routine that tends to inoculate us from new experiences and so inhibits our ability to explore. Our curiosity begins to slip away. A reasonable aim for the scientist is to try buck the trend of adult life, by maintaining a state of permanent curiosity like the one we all had when we were young. Indeed, the idea of the "scientist as a child," cheerfully playing with equipment in the lab like a youngster playing with a ball in the garden, is well known, almost to the point of being a cliché. But there is one big and obvious difference: Discoveries by children are personal revelations of things (like talking, walking and throwing a ball) that have already been discovered by others. In contrast, the scientist aims to discover things that nobody else knows. While it's true that a child-like sense of wonder goes a very, very long way in science (without one, you might as well quit), it is equally true that wonder alone is not enough. So what exactly does it take?

A Planetary Discovery

I am a planetary astronomer, which means that I use telescopes to examine objects in the solar system to try to figure out new things. My big-picture science motivation is to pin down how the solar system formed and how it has evolved since formation. It's pure science with no economic value, of use only in furthering human understanding of the world. On the other hand, it has immeasurable value for me as a person; research is one of the great pleasures of my existence. Around the world there are probably 100 or 150 planetary astronomers, with most of the interesting new work done by maybe 10 or 15 of them. The name of the game and the thing that distinguishes one astronomer from another, is to know what is worth measuring in the first place. In other words, the key is to ask a good question.

The textbook description of "the scientific method" is that theorists make predictions that observers try to refute. In truth, it's much more messy than that and the tables are often turned: observers make surprising discoveries that theoreticians struggle to capture in their models. Instead of being a deterministic exchange between predictive theory on one side and observational tests on the other, the advancement of science has a large random component. There is absolutely nothing wrong with this. Good science is often about surprises and moments of jaw-dropping astonishment, typically without forewarning from theorists. The way to encounter these surprises and astonishments is to look, loosely guided by an idea, using the best equipment you can find and with the most open mind you can muster. So in my own case, for example, I read the scientific literature voraciously and I certainly try to keep abreast of current ideas. But, in the end, I know that what matters most comes down to a telescope, to luck and to me.

In the 1980s, the inner solar system was known to be home to thousands of asteroids and comets as well as the planets. In contrast, the outer solar system had Uranus, Neptune, Pluto and not much else. This dichotomy between the "empty" and "full" parts of the solar system struck me as peculiar and unnatural. My then-student, Jane Luu, and I reasoned that the emptiness of the outer solar system might be real (because the distant giant planets might have been able to scatter away any nearby objects). Or it might be an artifact; objects viewed in reflected sunlight fade very rapidly (as the inverse fourth power) with increasing distance from the Sun. Perhaps the outer regions were full of objects too far away, too small and too faint to have been detected, given the technology of the day. We didn't know which one, if either, of these possibilities was correct, but the simple question "why is the outer solar system empty?" seemed like a good one. Others before us had asked this same question but did not answer it. We decided that we should try.

The only thing to do was to have a look. We set as our goal to find any new objects orbiting the Sun beyond Saturn (which is at 10 AU, or ten times more distant from the Sun than is the Earth). We explored several strategies, but soon settled on the use of CCDs (charge-coupled devices similar to the detectors embedded in modern, electronic cameras) to take multiple pictures of the sky through telescopes pointed in the direction opposite to the Sun. When compared visually on a computer, pictures of a given patch of sky would show moving objects jumping across the image relative to the (fixed) stars and galaxies in the background. In the direction opposite to the Sun, the speed of any moving object should be inversely related to its distance (for the same reason that a high-flying airplane appears to cross the sky much more slowly than a low altitude one, even if they both have the same speed). We sought distant objects, expected to have slow motions relative to the stars. Perhaps a little unimaginatively, we called our project the SMO ("Slow Moving Object") survey.

We found nothing of interest for over five years until, in 1992, we discovered an object orbiting far beyond anything else in the known solar system. The so-called 1992 QB1 had a nearly circular orbit some 43 AU from the sun (Neptune is at 30 AU) and was about 250 km in diameter (compared with the Earth's diameter of 13,000 km and the Moon's of 3500 km.) This is already larger than almost all asteroids in the main belt between Mars and Jupiter. In the following decade, we continued our survey and found many dozens of trans-Neptunian objects. Our success brought others into the search, triggering an unprecedented avalanche of discoveries in a region now variously called the Kuiper belt or the Edgeworth-Kuiper belt or, simply, the trans-Neptunian belt. At the time of writing, the belt has 1500 known members, and the projected population is several billion objects larger than a kilometer. For every ordinary asteroid in the main-belt there are roughly 1000 Kuiper belt objects beyond Neptune. We had discovered that the solar system beyond Neptune is teeming with primordial objects.

Video - Explanation of Kuiper Belt in 10 minutes - tinyurl.com/KuiperBelt10

This is not the place to describe in gory detail the scientific significance of the Kuiper belt. Suffice it to say that our discovery has revitalized and revolutionized the study of the solar system, particularly of its formation and evolution. Kuiper belt resolved the long-standing question of where short-period comets come from. It is a repository of the most primitive, least thermally processed matter in the solar system. Comets are aggregates of ice and rock that grew in the frigid outer reaches of our system. Most were ejected from the solar system in an early, chaotic phase, and are lost forever amongst the stars. Some of those that formed beyond Neptune, however, escaped being scattered out and have remained more or less where they formed ever since. The Kuiper belt is essentially a comet nursery. Moreover, temperatures beyond Neptune are so low (-230 Celsius and colder) that there is no chemistry; the original materials accreted 4.6 billion years ago have simply remained frozen there.

Our work showed that the motions of Kuiper belt objects fall into distinct groups. For example, the so-called "Classical Kuiper belt objects" (KBOs) have relatively circular orbits aligned in the plane of the solar system. I called them "Classical" because their orbits loosely resembled the classical picture of the primitive solar system that astronomers had developed over the years. Others move in special orbits that allow them to avoid Neptune at all times, even though their orbits sometimes cross that of the planet. I called these "Resonant KBOs" because, in technical language, they occupy "resonances" with Neptune that convey protection from ever meeting the planet. Pluto is a resonant KBO, so I labeled similar KBOs "Plutinos" ("little Plutos") as a tongue-in-cheek way to make the connection memorable. Still others follow vast, looping orbits that pass near Neptune but reach out dozens, hundreds or even thousands of AU. These "Scattered Kuiper belt objects" are in the gradual process of being launched towards the stars by the cumulative effect of repeated gravitational impulses from Neptune.

Some of the orbital groups give a dynamical record of the evolution of the solar system showing, for instance, that the orbits of the planets have changed in size since the planets formed. The resonant objects have been especially important in this regard. Planetary dynamicist Renu Malhotra (University of Arizona) showed that KBOs could have been trapped in these special orbits if Neptune had moved outwards from its formation location. Nobody has come up with a better explanation and this "planetary migration" is now a central feature of solar system understanding. Migration has upset the apple cart by destroying centuries-old ideas about a repetitive, clock-like solar system and replacing it with one in which seemingly crazy events might have happened in the distant past. For example, we now recognize that the planets might have experienced an unstable phase in which massive Jupiter and Saturn disturbed each other's orbits, throwing the whole solar system briefly into chaos. As the planetary orbits expanded, their strong gravitational perturbations would have disturbed the Kuiper belt, hurling billions of objects into interstellar space, never to be seen again. Others would have been launched towards the inner solar system, where many would eventually strike the Sun and planets. And Pluto, formerly the tiniest planet with the largest, most highly tilted and elongated orbit, is now understood as just a large but otherwise ordinary Kuiper belt object.

Video - Lecture by David Jewitt on The Significance of the Kuiper Belt - tinyurl.com/significancekuiper

Since its discovery in 1992, the Kuiper belt has been discussed in more than 1400 refereed scientific papers, giving some measure of its impact on solar system science. Given this, it is fun to consider some of the obstacles we faced on the way to finding the Kuiper belt.

First, we had a very ill-defined target. When we started our survey we didn't know where to expect objects, or how many there might be, or how big or bright, or even if they existed. Our limited aim was simply to find "anything more distant than Saturn," on the understanding that any such object would be interesting because of its uniqueness. Years into the survey, several computational dynamicists wrote papers suggesting that, while the region inside Neptune should have been clear of objects, the space outside might not be. These papers provided moral support for our search, but did not motivate it.

Second, when we started in 1986, we didn't know that the equipment at our disposal was inadequate to the task. Our first telescope was too small and its detector too puny to show the trans-Neptunians. This is obvious in retrospect, of course, but we had no idea at the time. If we had known in the beginning that our instrumentation was incapable of doing the job, we wouldn't have started. Sometimes, ignorance is bliss.

Third, access to telescopes is allocated in a way that does not necessarily encourage discoveries like ours. Observing time is precious and is competitively awarded by impartial committees of other astronomers called TACs (Time Allocation Committees). The telescopes are over-subscribed and many TAC-members want to use the same telescopes for their own projects, so that I sometimes wonder if they are as impartial as they are supposed to be. Even without obvious conflicts, most pre-Kuiper belt astronomers had little interest in the solar system - it was a back-water not a "hot topic" in their minds compared to exciting developments in extragalactic astronomy. That's understandable because, when we started, there was nothing in the outer solar system to be interested in. The TACs at first awarded us less telescope time than we requested, then no time, accompanied by discouraging (always anonymous) comments.

Despite the negativity, the absence of an effective telescope policing system gave us room in which to maneuver. For example, when our telescope proposals were rejected, we responded by requesting observing time for more palatable (i.e. routine) science projects. When allocated, we used the time to do the SMO survey as we wanted. There was nobody on-hand to check what we actually did with the telescope.

Fourth, NASA is the main American source of funding for planetary science research, but it is not well set-up to encourage our type of discovery. NASA review panels, like TACs, are composed of "peer scientists." Like TAC members, panelists struggle to be impartial because they feed from the same trough. They also tend towards the conservative. NASA panels find it easier to support "incremental science," in which the path to the end is more clear and the result can be confidently predicted. This is not wholly unreasonable; a lot of public money could be wasted on scientific wild-goose chases. To fund a search that had produced nothing, and which might very well continue to produce nothing, evidently seemed to my colleagues on the NASA review panels to be a step too far. My proposals for grant money to support the SMO survey (for example to pay Jane, other students and part of my own salary) were rejected. Out of necessity, I used money allocated for other "incremental science" projects to support the SMO survey. In this sense, the Kuiper belt was discovered despite NASA's best intentions, not because of them.

Any one of these problems (the absence of predictive models, uncertainty about what we were looking for, inadequate equipment, denial of telescope access, denial of funding), could have killed our search for slow moving objects. Even if these hurdles had not existed, there's no law of science that requires the distant regions of the solar system to be occupied. Our system could happily exist without a Kuiper belt and we might have found nothing at all, no matter how hard we had tried. So, the final ingredient contributing to the success of our discovery was “good fortune." We were simply lucky that nature gave the solar system a heavily-populated trans-Neptunian space; it could just as easily have been empty.

People ask why we kept going with the Slow-Moving Object survey, for such a long time without result. The main reason was that we very much liked our own idea; we thought that our question about the emptiness of the outer regions was so simple that it deserved an answer. We knew that “absence of evidence is not evidence of absence” and that “out of sight is out of mind." And we felt that this was a promising subject precisely because other people were not working on it, at least when we started. We also knew that our cameras and computers were steadily improving, a benefit of Moore’s Law (according to which the speed of electronics doubles every 18 months or so). As new equipment became available we were able to quickly re-do all the work we had done before and then to surpass it in a fraction of the time. Lastly, just as it might be tempting to abandon a long wait for a bus, we had a nagging (and probably irrational) suspicion that the bus was just about to arrive. We simply didn’t want to give up too soon.

Video - Lecture on solar system science: Planets and Exoplanets - tinyurl.com/planetsexoplanets

Notes:

My web site includes a general background on Kuiper belt - http://www2.ess.ucla.edu/~jewitt

I’ve described the discovery of the Kuiper belt elsewhere (D. Jewitt, The Discovery of the Kuiper Belt. Astronomy Beat, Astronomical Society of the Pacific, 2010).

Asymmetry of Brain Function

Lesley J. Rogers

Lesley Rogers is Emeritus Professor, University of New England, Armidale, NSW, Australia, and Honorary Professor at Queensland Brain Institute, Brisbane, QLD, Australia. She was elected as a Fellow of the Australian Academy of Science for her outstanding contributions to understanding brain development and behaviour. She has published 16 books and well over 200 research papers.

1. Discovering that not only humans have lateralized brains

In the 1970s I was conducting research on the molecular processes that take place when memories are laid down in the brain. I was blocking memory formation in young, domestic chicks (*Gallus gallus*) by treating their brains with drugs known to inhibit specific cellular processes thought to be essential for memory formation. One of these drugs was the inhibitor of protein synthesis, cycloheximide, chosen because long-term memory is thought to depend on nerve cells making new proteins. I discovered that cycloheximide does block long-term memory but that is not all it does. It also causes long-lasting impairment of further learning that depends on both long- and short-term memory.

Following treatment with cycloheximide two days after hatching, chicks are very slow to learn to find grains scattered amongst small pebbles when they are tested days, weeks or even many months later. It was not an effect that I had expected but one worth studying further because researchers at that time, using protein synthesis to block the formation of memory in other species, thought it had a specific effect on memory only.

At this time my colleague and mentor, Professor Richard Mark, then at Monash University in Australia, suggested that I try administering cycloheximide to just the left hemisphere or the right hemisphere of the brain to see whether this gave different results. He had thought of this because he had spent time in the laboratory of Roger Sperry, the famous researcher, and Nobel Prize winner, who studied human patients who had had the large nerve tract connecting their left and right hemispheres, called the corpus callosum, sectioned in an effort to stop debilitating epilepsy. They are known as "split-brain patients." Sperry studied split-brain patients to see what functions are located in the left and right hemispheres in humans.

At that time and a hundred years prior to it, it was believed that only humans have different specializations of the left and right hemispheres of their brain. Since language is a function of the left hemisphere in most humans and we are predominantly right-handed, this was thought to be a quality separating us from all other species. Earlier observations of hand preference in primates (apes and some species of monkeys) had failed to find evidence of hand preferences and, consequently, it seemed to be a closed issue about human superiority and uniqueness. No one considered looking for lateralization in the brains of non-primate animals.

It seemed simple to test the ability of chicks to learn after treating either their left or right hemisphere with cycloheximide. To my surprise, I found that, although treatment of the chick's left hemisphere with this inhibitor caused long-lasting impairment of learning, the same treatment of the right hemisphere had no effect. We had discovered lateralization in a non-human brain!

At about the same time, Fernando Nottebohm, at the Rockerfeller University, found that canaries control production of song using neural circuits in their left hemisphere and not in their right hemisphere. He showed this by placing lesions in the left or right higher vocal centre in the brain hemispheres and he found that only left-sided lesions prevented the canaries from singing. Right-side lesions had no effect on singing. His research was published, with colleagues, in 1976, and mine in 1979. I was not aware of Nottebohm's research when I was conducting my experiments on chicks. His finding subsequently attracted more public attention than mine, probably because people saw a connection between birdsong and language and were more inclined to accept that they were both functions of the left hemisphere. However, our separate investigations provided the first evidence to overthrow the idea that only humans have brain lateralization (hemispheric specialization). As it turned out, belief in the uniqueness of brain lateralization in humans was not about to be cast aside lightly.

Once I had made the discovery in the peaceful surroundings of my own laboratory with supportive colleagues, it was time to tell others. It was then that I realized what opposition the finding would receive. Some colleagues just did not believe it. Others, especially those working on humans, were incensed by it. They were reluctant to give up their view of laterality being special and uniquely associated with language; some held on to this view for decades, even long after lateralized brain function had been shown in a range of vertebrate species.

It was this response that led me to engage more strongly with research on laterality. At first I went about documenting more functions of the chick brain that are controlled by one or the other hemisphere. Later, as the field expanded and more researchers began to see that lateralization is present in animals, I also investigated lateralized behaviour in amphibians, mammals (including primates), and other avian species. In fact, research on lateralization has captivated my interest for the last three or four decades. I am still working on it, now in bees, by looking at memory recall in bees trained to associate a sugar reward with a particular odour (e.g. lemon) and not with another odour (e.g. vanilla). Even invertebrates have lateralized brains. When bees are trained using both antennae to sense the odours and then tested using only their right antenna, they can recall the memory but only up to about 3 hours after training. When they can use only their left antenna in the recall tests, they cannot retrieve the memory for the first 3 or so hours after training but by 6 hours after training and on the next day they can recall the memory only when using their left antenna. Access to the memory is lateralized and changes with time.

Demonstrating lateralized behaviour in chicks was made simpler, and more humane than injecting drugs, after Professor Richard Andrew, at the University of Sussex and formerly my PhD supervisor, discovered that simply covering one or the other eye revealed different behaviour. When the chick has its left eye covered, and so is using its right eye and left hemisphere, it can learn to find grains scattered on a background of small pebbles. It cannot do this when its right eye is covered. This method of showing lateralization of visual behaviour is possible only in species with their eyes placed on opposite sides of their head because, in them, each eye sends most of its input to the opposite side of the brain.

Of course, once discovered, it was important to find out why brains are lateralized. Although having a lateralized brain might have been essential for the evolution of language in humans, there must have been much earlier, and other, reasons for having a lateralized brain. In a wide range of vertebrates, the left hemisphere is used when learnt behaviour is performed and the animal needs to focus on what it is doing, as in the pursuit of prey. The right hemisphere is used when the animal has to respond to emergencies, as in recognizing and escaping from predators, in attacking and in expressing fear.

Much more research will have to be done before we know why lateralized brains evolved but we have some convincing evidence that chicks with lateralized brains perform better when they need to look out for predators at the same time as they are searching for food. They apply the left hemisphere in searching for food and the right hemisphere in detecting a predator. Separation of functions that require different modes of processing appears to be an advantage. This discovery was made when I worked with a visiting researcher from Italy, Dr Paulo Zucca, who came to my laboratory in 2003. We published it in 2004 together with Dr Zucca's supervisor and my long-standing colleague Professor Giorgio Vallortigara.

We made this discovery by testing chicks in their second week of life on the task of searching for grain scattered amongst small pebbles and at the same time we moved the silhouette of a hawk overhead. We scored whether or not they detected the overhead 'predator' and how many mistakes (pecks at pebbles) they made. We compared chicks with and without lateralization for these functions.

Lateralized chicks were able to perform these two tasks at the same time but non-lateralized chicks had great difficulties. The non-lateralized chicks either missed seeing the predator or they were so disturbed by it that they made more and more mistakes in pecking.

Lesley J. Rogers

2. Light effects on the development of lateralization in avian brain

At the time when I first discovered lateralization in the chick brain, most of my colleagues said that genes must control its development. It was the beginning of the time when scientists were being swayed towards genetic determinism of behaviour, not only in animals but also in humans. I saw this unitary approach as incorrect and it stirred me to try to think of a way in which lateralization of the chick brain might be influenced by some factor in the environment.

One day I was looking at a series of pictures of chick embryos and I noticed with a start that, just before hatching, the embryo is oriented in the egg with its head turned to rest against the left side of the body, thereby occluding the left eye and leaving the right eye next to the membranes and shell where it can be stimulated by light. Perhaps, I thought with excitement, this asymmetrical stimulation of the eyes had contributed to the development of hemispheric differences in processing visual information.

The next step was to test this. My hypothesis was that chicks hatched from eggs incubated in the dark during the final few days before hatching would not be lateralized when I tested them on the task of finding grain scattered amongst pebbles (no left hemisphere superiority in performing this task) and they would not show heightened attack behaviour when they used their left eye and right hemisphere. This turned out to be correct. The results were published in *Nature* in 1982.

Although turning of the embryo's head may be controlled by genes, the generation of visual asymmetry depends on exposure to light. In fact, I was later able to show that visual lateralization could be reversed by removing one end of the egg shell some three days before hatching, gently easing the embryo's head from the egg and applying a patch to the right eye while allowing the left eye to be stimulated by light. After hatching these chicks used their right, and not their left, hemisphere to find grains scattered amongst pebbles and they attacked when they used their left hemisphere. Their brain function had been reversed. This provided totally convincing evidence of the role of light in establishing the lateralization.

3. Processes of research

I think it will be clear that these discoveries were made according to steps that involved serendipity, intuition, and well-planned design. Break through on at least two occasions was made by thinking outside accepted beliefs and, of course, by passionate pursuit of the answers.

Discovery of asymmetry in the avian brain was the beginning of a new area of research. There are now many researchers investigating lateralized behaviour and brain anatomy and chemistry in many different species. Out of this some model species for studying the development of lateralization have emerged and they are now being used to study genetic and epigenetic influences important in the development of lateralization. Recently, the potential of research on laterality in animals to shed light on the processes involved in abnormalities of lateralization in humans has been realized. Weaker lateralization is known to occur in humans with depression, schizophrenia, autism, and stuttering. Despite the large amount of research on lateralization in humans, precise evidence linking brain asymmetry to behaviour will only be found by conducting research on animals because, using animals, we can do well controlled experiments and we can investigate the dynamics of development. Such experiments cannot be performed on humans.

Of course, to come back to the time when I began my research on this topic, the first steps showing the error of believing that lateralization is unique to humans were exciting and paradigm shifting.

Throughout my research was funded by the Australian Research Council.

Interview with Lesley Rogers - https://www.science.org.au/node/328009

References that reported these discoveries

Rogers, L.J. and Anson, J.M. (1979) Lateralisation of function in the chicken fore-brain. *Pharmacology, Biochemistry and Behavior*, 10, 679-686 (1979). [Anson was my student at the time.]

Rogers, L.J. (1982) Light experience and asymmetry of brain function in chickens. *Nature*, 297, 223-225 (1982).

Rogers, L.J. Light input and the reversal of functional lateralization in the chicken brain. *Behavioural Brain Research*, 38, 211-221 (1990).

Rogers, L.J., Zucca, P. and Vallortigara, G. (2004) Advantage of having a lateralized brain. *Proceedings of the Royal Society London B*, 271, S420-S422.

Rogers, L.J. and Vallortigara, G. (2008) From antenna to antenna: Lateral shift of olfactory memory recall by honeybees. PLoS ONE 3(6): e2340. doi:10.1371/journal.pone.0002340. www.plosone.org/doi/pone.0002340.

Recent summary book - *Rogers, L.J., Vallortigara, G. and Andrew, R.J. (2013) Divided Brains: The Biology and Behaviour of Brain Asymmetries. Cambridge University Press.*

The Journey is the Point

Sheryl L. Bishop

Sheryl L. Bishop, PhD is a professor in the Schools of Nursing and the Graduate School of Biomedical Sciences at the University of Texas Medical Branch at Galveston, Texas. In addition, Sheryl has served as a faculty for the various programs at the International Space University (ISU) since 1996. As an internationally recognized behavioral researcher in extreme environments, for the last 25 years Dr. Bishop has investigated human performance and group dynamics in teams in extreme, unusual environments, involving deep cavers, mountain climbers, desert survival groups, polar expeditioners, Antarctic winter-over groups and various simulations of isolated, confined environments for space, including a number of missions at remote habitats (e.g., Mars Desert Research Station, Utah, and FMARS and the Mars Project on Devon Island, Canada). She has been a reviewer for multiple scholarly journals as well as a grant reviewer for the European Space Agency's Concordia Station, the Canadian Space Agency's Life Science Directorate, the Australian Antarctic Science Division, and the Czech Science Foundation. She routinely presents her research at numerous scientific conferences, has over 60 publications (including contribution to NASA's Historical Series on Psychology in Space) in both the medical and psychological fields. She is frequently sought out as a content expert by various media and has participated in several television documentaries on space and extreme environments by Discovery Channel, BBC, the History Channel and 60 Minutes.

They say that life is a journey. In looking back at mine, it's been quite a trek.

From the perspective of "now," where I sit comfortably with a modest international reputation and recognition as one of the handful of psychologists and behavioral scientists that have spent 25+ years pursuing the answers to selecting the "best" individuals crazy enough to want to go into space but sane enough to do so (to paraphrase Kim Stanley Robinson's psychologist in *The Martians,* 1999), back to the beginning of becoming who I am today, it's been quite an adventure.

Now having an adventure in my family was not always an entirely enjoyable thing. In trying to focus on the upside of challenging situations, my mother would try to spin a particularly disastrous camping trip or roadside breakdown as an "adventure". In a family of five children, our exploits frequently involved trips to emergency rooms or lots of practice resisting the urge to inflict bodily harm on particular others. In the end, the experience served us well. Coupled with vast reservoirs of humor, these adventures became part of our family mythology and provided the mental scaffolding for our development as adults.

Hence, in the mid-70's I was trying to settle on a career in order to finish the bachelor's degree that I'd been leisurely working on for way too long. As an avid consumer of science fiction since early childhood, I decided to visit Johnson Space Center in Houston, Texas, to see what career opportunities were available. I wanted a career that satisfied my generalist, jack-of-many-trades personality. They advised me to look into closed loop environments like engineering. As engineers were plentiful at that time, I decided to project a bit further into the future and it occurred to me that there would come a time when the need to address the myriad of issues surrounding human adaptation to the space environment would become important. So I settled on social psychology with a specialty focus on space. However, making the decision and setting foot on the path, in and of itself, was a challenge.

Having located a desirable mentor in the late Robert Helmreich at the University of Texas in Austin, I needed to convince the powers that be in a doctoral program noted for its challenging curriculum to take a chance on an older, returning student with a family in tow. Living an hour and a half from Austin, I enrolled as a special student and attended classes throughout the last tri-semester of my pregnancy with my second child. Whether it was admiration at such tenacity or nervousness that one of my instructors would have to deliver my child, halfway through the semester, I was told to "go home, have your baby….We'll admit you in the fall."

So off I went to graduate school ready to study human adaptation to space. To my chagrin, I found access to actual space crews nonexistent and, once again, realizing that there was work that needed to be done before we would ever have large numbers of actual crew members in space, I sought the next best thing: analog environments. Since any analog to space needs to mirror some aspects of confinement, isolation, and threat, this lead me to the various extreme environments I've studied over the last 25 years.

Little did I realize that a decision to study the "next best thing" to space, i.e., terrestrial analogs, would lead me into the Australian outback in full camouflage, surreptitiously following teams of males and females as they walked across the desert foraging off the land for three days. Or find me on the side of Mt. Aconcagua at 8,000 feet keeling off my horse with altitude sickness on my way up to base camp at 11,000 feet. Or having to answer the exasperation of my Greenland glacier expeditioners as they explained how my plastic questionnaire binders broke in the extreme cold and they had to chase the pages across the icy glacier. And so on…to so many, many adventures that have informed the process of discovery.

At the heart of space exploration is the assumption of an off-world human presence. Regardless of whether we make our first commitment via robotics or lead the way with human explorers, the central premise is that humans will be an intricate part of the expansion into space. Of necessity, the inclusion of humans magnifies the complexity of any mission exponentially, whether it be off-world or merely to a remote location on-world. Humans are both the strongest and the weakest links in a system. Physically more fragile than mechanical counterparts, we require very complicated systems for life support. The determination of these life support systems are not purely driven by objective parameters. For instance, early discussions about hygiene systems on Space Station Freedom between Russian and American design teams came to a dead halt when unresolvable differences in preference emerged between a shower system and a sauna system. Similarly, insistence on different water treatment approaches (iodine versus silver nitrate) could not be bridged. In the end, International Space Station (the inheritor of design decisions for

Freedom) had neither a full shower or sauna and two separate water treatment systems for the different crews.

If that were not enough, being human means there are also demands and requirements for psychological well-being that need to be addressed. The inclusion of humans means nothing can be truly standardized nor specified with much certainty. There are no exact specifications that can be imposed upon design requirements, no template from which to manufacture exchangeable and replaceable parts, no operating parameters that will limit function or define nominal performance. In short, the inclusion of humans in mission profiles gives engineering types heartburn.

I have on many occasions reminded my engineering colleagues that an engineer could build the most marvelous hardware, design the most amazing software, and then the 'wetware' could be counted on to come along and screw it all up. So why would we go to all the trouble to include such fragile, fallible, and unpredictable elements in an endeavor costing more than some national GNPs? *Because space is unpredictable.* Machines lack the capacity for creative problem solving and adaptive responses to unpredictable, dynamic, and unforeseen situations. For that, you need the capacity to exceed known parameters, to apply principles in untested and newly envisioned ways, to step outside programmed operating parameters and evolve. For that you need humans.

Given the need, rationale, and pure emotional appeal of including humans in exploration, the challenges have been approached in various ways in human history. Prior to the modern era of risk aversion, expectations of attrition and crew loss in exploration were simply parts of the equation of expedition planning. Party numbers were calculated with loss projections as part of the logistics as pragmatically as calculating the amount of rations needed per person. Risk was understood on both sides of the endeavor--by those that were going and those that were staying behind. Thus, the qualities sought were experience with the territory to be explored, skills pertinent to the environment, and a hardy physique likely to survive the rigors expected. Little, if any, attention was given to psychological fitness although clear evidence of mental illness was generally disqualifying under medical considerations.

So what makes space different? One might argue that the selection of crews for space wasn't different initially. We took the most highly skilled of those in the aviation field--test pilots--not for their flying skills per se since everything was automated and controlled from the ground, but for their, hopefully, analogous inoculating experience with various high risk forms of flight that would better enable them to deal with the mental challenges of space flight with its vast number of unknowns. In selecting military aviators, we could check off the experience box (or as close as we could get to it), the physical fitness box, and the mental toughness box. The skills pertinent to the environment were a huge unknown against which human capability was viewed with both reassurance (flexibility, adaptability, real-time problem-solving) and anxiety (mental breakdowns, physiological and psychological dysfunction). We simply didn't know enough about space to know what to worry about and what we didn't have to be concerned about. So we worried about everything.

The microgravity environment turned out not to be filled with many of the monsters our imaginations conjured ("here there be dragons"). We have made great in-roads on a huge number of technological challenges (e.g., efficient life support systems, radiation shielding), medical challenges (e.g., bone and muscle degradations, cardiovascular impacts, orthostatic disruptions), and human factor issues (e.g., work stations that didn't disorient in a 3-D microgravity environment, ergonomic tools, and suits for performance under pressurization). If it was quantifiable, there was an avid dedication to finding answers. Unsurprisingly, focus on the "human" side of the equation lagged since quantifying interpersonal team dynamics exceeded all engineering models. Human issues were not particularly amenable to engineering fixes. Yet, there were issues.

Early approaches took the stance of unrealistic expectations that human crew members could (and would) suspend all that messy humanness and function "professionally" with machine-like efficiency and effectiveness. For very short periods of time, under ideal conditions, such adherence to machine-like task centeredness was possible. But, eventually, inevitably, the human side would make its appearance. This was succinctly driven home in the early 1990's when an astronaut quietly slipped into a meeting of psychological and behavioral researchers gathered to review the status of psychological knowledge and research needed for space missions. Deferentially stepping forward, he wanted to be sure the group knew that (1) not all astronauts viewed psychologists negatively (i.e., who were only marginally less likely to be voluntarily sought out than a flight surgeon), (2) that many of the astronauts were absolutely aware and convinced that our concerns were right on the mark and something needed to be done, and (3) just to reinforce that point, he related that on a recent mission, "If it had gone a single day longer, there might have been one less person for the return trip." To doubly reinforce that message, he shared the fact that every single crew member had *voluntarily* sought out the NASA psychologist after the flight to discuss the situation. Consider us impressed with his willingness to share, but he was preaching to the choir.

You see, most of the astronaut corps also bought into the official dictate that professionals could put aside human emotions for extended periods of time and stay focused on the task at hand. While they did, indeed, continue to successfully 'get the job done', as missions grew in length, in number of crew members, in diversity of professional, national and gender backgrounds and became more mundane and routine with less public scrutiny, less fame, less hype, less importance…less of all those factors that keep the adrenaline pumping and cause individuals to be on their best behavior, the potential impact of the interpersonal side of being human became greater.

The group of astronauts, while still an elite, select few, was not the handful of individuals with mirror backgrounds (i.e., test pilots) that would naturally form a cohesive in-group. In addition, as the corps grew larger, opportunities for being chosen to fly became more competitive. Until you actually flew a mission, you were only an astronaut candidate in the corps. The golden ring was being selected for a mission. You simply didn't get to be good enough to be selected as an astronaut candidate without being vested in being the best of your field. The drive to be the best took on various forms - competitiveness, perfectionism, achievement motivation, drive to succeed, and/or exceptional persistence, endurance and dedication. But these are human drives and needs and not the kind to relegate to passionless, robotic dedication to external organizational end-goals. They *cared.*

And therein lay the field for conflict. Realization that all that psychological 'stuff' really did and was going to continue to be a big deal for future space missions was almost resisted as strongly by the members of the corps as it was by the agency. But the members of the corps were living with it, grappling with the interpersonal differences on a daily basis. When international missions began, the merely heretofore 'troublesome' professional and gender differences were exponentially increased with the introduction of different nationalities, cultures and space agency operational approaches. When you have a crew member who refuses to even acknowledge verbal directives by a commander because she is female and his culture doesn't encompass taking orders from a woman, it doesn't take long for the realization to sink in that something needs to be done. Even among homogeneous gender and national crews, quiet, suppressed instances of conflict and discord could be found.

In 1993, Santy et al interviewed the majority of astronauts that had flown at that time on the types of issues that provoked conflict and discord and had them rate these as to mission impact (Santy, Holland, Looper & Marcondes-North, 1993). Surprisingly, it was mundane matters of routine housekeeping and personal hygiene that accounted for the greatest number of incidents as well as those most significantly impacting mission functioning. The prospect that who did or who didn't do the dishes could become a serious mission threat wasn't even on the radar of those involved in taking the next big step forward off Earth. But it should have been.

It took a while longer for awareness and acceptance of the importance of interpersonal team dynamics to become a serious part of mission planning and focus. It was purely human responses that underlay the infamous Skylab Mutiny (1973) when an exhausted four-man rookie crew took matters into their own hands and took an unscheduled and unsanctioned day off to rest aboard the Skylab space station; or the expressive outpouring of relief and joy by Norm Thagard over the end of his cultural isolation as fellow Americans arrived at MIR to return him home; or the weight of the growing evidence by astronauts themselves of interpersonal friction between team members that impaired or impacted mission functions (Kanas, 1987; Santy et al, 1993). Similar evidence was being amassed in analog environments as well by myself and others (Palinkas, 1991; Bishop, 2002; Bishop and Primeau, 2002; Bishop, Grobler & SchjØll , 2001; Sandal, 2001). It soon became clear that, no matter the environment and despite individual skill and experience, factors such as confinement, isolation, and high risk took their toll on participants.

One of my earliest studies followed the team members of the Huautla Deep Caving expedition over nearly three weeks of exploration of the world's then 12th deepest cave system in southern Mexico (Bishop, Santy & Faulk, 1999). Their saga encompassed a leader whose single-minded focus on making a break-through discovery into uncharted territory most certainly contributed to the exhaustion and over-exertion leading to the death of a team member. This leadership drivenness, without doubt, contributed to the effective mutiny and departure of all but three of the original 16 team members over the course of the expedition. Yet that same drivenness was also responsible for the remarkable breakthrough discoveries of new territory that moved Huautla from 12th to 5th deepest cave at the end of the expedition.

The team of assembled 15 international cave divers trained for over two years for the grueling expedition. The accidental diving death of a team member during training foreshadowed the challenges they faced. In addition to a number of support members, a National Geographic film crew accompanied the team the first few weeks as they worked their way to the designated base camp. There were two insulin dependent diabetics (one support member and one diver) on the expedition.

The mission was driven by the fact that progress from 18 prior expeditions exploring the extensive cave system had been virtually halted at a flooded underwater passage (called a sump) that stymied all previous attempts to find a way through. Located over a mile underground, the amount of scuba gear needed to bypass the various underwater tunnels to even reach Sump 5 made traditional diving gear unrealistic. As a member of the previous expedition, the current expedition leader had spent two years developing a new rebreather diving unit which did not require the traditional tanks of compressed air.

The expedition spent 2.5 months on location with 44 days in darkness within the cave system. Working in total darkness with only their headlamps for illumination impacted circadian rhythms and abilities to sleep. There was constant dampness and frequent prolonged periods of working in cold water (64° F) as they traversed passages that ranged from narrow, body-width tunnels to expansive rooms with waterfalls. In such rooms, the noise from falling water could range louder than 100db (as loud as a power lawn mower). The effort to climb, dive, swim, and crawl through the terrain in such an environment represented a constant danger of injury or death from slips, falls and drowning. Coupled with the extreme physical exertion and darkness, there was the unremitting awareness of isolation and remote access to help. It was not an environment for the faint of heart.

However, even those who had trained for the contingencies of this expedition, who had the requisite skills and experience in caving and diving, who were physiologically and mentally prepared for the rigors that lay before them…even these reached their limit. Group dynamics began to fray under a demanding schedule set by the expedition leader whose primary focus was on breaking through Sump 5. As he pushed team members to make multiple trips into the cave system hauling supplies and pushing forward to establish the base camp at Sump 5 and begin exploration, small groups began to rebel and take days of rest or short expeditions to nearby caves for a break. Eventually, even the team doctor insisted on a break and returned to the surface. On day 16, an exhausted diver, the insulin dependent diabetic made a second dive and experienced an apparent hypoglycemic blackout while diving and asphyxiated. The death highlighted the conflicting priorities between the leader and team members and contributed to the dissolution of the team over time. At expedition end, the team was composed of only three divers and one support member (all film team members had left according to schedule before the drowning). The cost to long-term personal and professional relationships, financial losses, reputations and other intangible dimensions is incalculable.

Yet the mission was officially a success. For even after the majority of his team had left, the expedition leader and another female team member with the team doctor and one other support member, reentered the cave, dove Sump 5 and found that elusive passage beyond. The known length of Huautla was extended from 3.3 km (2.05 miles) to 56 km (34.78 miles). The new depth of 1475 meters (4819 feet) moved Huautla from 12th deepest in the world to 5th for a while (it has since been displaced by new records in other caves). The successful traverse of 655 meters (2140 feet) of underwater tunnels and the difficulty of access classified Huautla as one of the world's most difficult cave systems; a designation that it holds even to this day (Steele, 2009).

My own role in the expedition was purely that of a bystander. Having met with the team in Texas before they caravanned to Oaxaca, Mexico, I had a brief opportunity to meet with team members, explain my study, distribute the journals and questionnaire booklets they would be filling out each day and gather my baseline measures. At the beginning of expeditions, all groups are characterized by high spirits, excitement, hopes, expectations, plans and high levels of energy. I would not be meeting with the team members again since most were returning home from Mexico. Their study booklets were to be gathered together and returned to me by expedition support personnel. All the baseline measures (psychological tests) were sealed and would remain that way until the expedition and study data collection was complete for purposes of confidentiality and anonymity.

Communication about the expedition progress or problems was through an expedition staff member in New York. Thus, it was not until the delivery of the journals that I became aware of the drama surrounding the drowning and dissolution of the team. I scrambled to reach team members and debrief them to find out what had happened. As I was to learn in the subsequent years of studying teams in extreme environments, such environments imposed an alternate reality upon those who inhabit them. The very rules for prioritizing things become anchored to the factors of the situation instead of more external drivers. Cries of 'you don't understand what we are dealing with' are a frequent refrain from remote crews. Winnowing 'truth' from perception in extreme environments is complicated by the fact that what is true for a group in an isolated, confined and extreme environment may not apply anywhere else. So how can those of us not *there*, decide what was true?

From the Huautla Expedition journals and debriefs, it was apparent that the leader was perceived as both the hero and the villain of the tale. The lines for personal responsibility grow very blurred under the intense pressure of interdependencies that form in small groups operating in very dangerous environments. The opportunity to engage with others on the surface who were not formally part of the expedition (numerous guest divers and cavers visited the site over the expedition period) certainly contributed to the fractionation among members and very likely to the heightened feelings of frustration and subsequent drivenness of the leader. On the other hand, once underground, the desire to assuage demands for greater performance very likely played a significant role in the decision to make that ill-advised second dive for the diabetic diver. Exhaustion plays no favorites and cares nothing for gestures of loyalty and support.

The lessons of Huautla are not lost. In fact, leadership has been clearly shown to be one of the key defining factors of successful and unsuccessful teams, depending on your definition of success. Yet what makes a strong leader a positive or negative contributing factor to mission outcomes is still largely unresolved. The very different outcomes of the famous competition to reach the South Pole by Roald Amundsen and Robert F. Scott were largely due to the differences in the leadership style of the two men. The Huautla study was my early introduction to the difference one person can make in an extreme environment that has little tolerance for misjudgment. It was also my first wake-up call to the short-comings of remote measurement; a theme that would play out in years to come in many ways.

Over the course of the next twenty-five years, it became increasingly obvious to the handful of psychologists, psychiatrists, and behavioral researchers that stubbornly persisted in the field that predicting who was best fit was not going to be a straight-forward task. The person-environment fit as well as the person-person fit had to be considered. The interpersonal dynamics within a team could mitigate or exacerbate situational conditions and vice versa. Identifying which factors were critical and which were merely moderators was a daunting task. Every factor was up for examination and some have been slow to be addressed due to reluctance to change the status quo. For instance, was gender or ethnic composition a critical factor? Concerns about mixed gendered teams delayed the inclusion of women until Shuttle was operational. Yet, if anything, mixed gendered teams have been found to be better than all male or all female teams in both analogs and space crews.

It was evident that such concerns were as much a reflection of cultural attitudes as objective concerns. Even early evidence in the behavioral sciences that suggested that women tolerated close, confined spaces more easily than men (Freedman, 1975) was largely ignored. Similarly, evidence that women may physiologically actually tolerate low gravity environments better as the orthostatic fluid shift is more easily compensated for in female systems designed to handle such shifts on a monthly basis (Fortney, Beckett, Carpenter, et al., 1988) was dismissed initially. Yet fears of sexual discord and distress sprinkled with puritanical anxieties regarding both genders living closely in isolation made it difficult to objectively evaluate mixed gendered teams against single gendered teams. As serendipitous opportunities to evaluate male versus female teams occurred, it was clear that while very different interpersonal styles of interaction certainly did exist, highly effective teams were equally likely in both genders.

This was exemplified by my field study of separate gendered teams crossing the Australian outback in 2001 (Bishop, 2001). The popular television show, Survivor, was scheduled to be filmed in Australia the next season and 60 Minutes Australia decided to do a special on a 'real' survival group. Contacting one of Australia's own noted survival specialists, Bob Cooper, they found that Bob had a group of individuals signed up for survival training fortuitously composed of six males and six females. The idea of a comparative documentary was born. An internet search on research into men's versus women's performance brought the producer to my door back in Texas. As I described the paucity of data on women's teams, the prospect of actually conducting a study while making the documentary became apparent to both of us. So off I went to Australia with my personality, stress, coping and team dynamic questionnaires and salivary collection tubes to enable us to also collect physiological biological indicators of stress, e.g., elevations of cortisol or DHEAS which respond to both physiological and psychological stress.

Bob Cooper met me at the airport in Perth with a Western Brown Snake, #2 on the list of Australia's most deadly snakes, in a trash can in the back of the car. Apparently he had been called to rescue and relocate the intruding reptile from the back yard of a local resident. My daughter's parting recitation of all the deadly critters inhabiting Australia flashing through my mind, I was fascinated. Meeting with the survival participants was no less interesting. A mixed group composed of a male nurse, a minister and his wife, three conservation workers, an air force aviation mechanic, a rigger, an engineer, and a secretary composed the general group. Ages for women ran from 25-41 and for men from 25-50. One woman was legally blind and in Australia able to participate in the Special Olympics. Her field of vision in bright daylight was about 30% and she had no night vision at all. Given that the teams would be up and walking well before daybreak and then into the early evening at times, her limited vision represented an extraordinary burden for the women's group.

A two day drive to the Pilbara in Western Australia brought us into outback country where sand goannas (big, 4.5 foot lizards) and the various species of kangaroos abound. Although it quickly became obvious that the 'roos' were as common as Texas jack rabbits, I still exclaimed with excitement every time I saw one to the vast amusement of my local team members. Mid-week found us at a 'station', a local ranch home typically situated on vast acreage, for a three day refresher course for the participants before they were to begin their 6 day trek across the Pilbra on their survival trek living off the land. During the refresher course, I briefed the team members on how to fill out the questionnaires and take the saliva samples. The 60-Minute crew consisted of a same gender reporter and camera person for each group. Each evening, the two survival instructors and I would meet the camera persons away from the team camps to resupply with batteries, film and food since they would not have time to forage like the teams. The reporters, on the other hand, were expected to participate as full team members and eat what they could find along with the others. For safety and logistic reasons, the daily check points for each team were in close proximity but not identical. This allowed us to follow the progress of both teams while remaining completely out of sight to preserve the sense of isolation and remoteness. In practical terms this meant the two instructors and I spent long hours driving our lorries over desert territory at night to get ahead of the teams and then trail behind at the point they passed us. My bunk was a bedroll on the back seat of the jeep in order to avoid wildlife snuggling up to me in the chilly desert night. Daytime temperatures were often in the high 90's to 100 F. Since the teams camped at mid-day and nights at the local waterholes, the instructors and I made do with snatched dips in the occasional stream, outdoor showers at station outpost camps or a waterhole shared with local cattle. It is remarkable how blissful such simple things as having enough disposable water available to wash the sweat and dust off can be.

Although both groups followed the same route for the most part, after Day 2, differential progress separated the groups such that they were not always at the checkpoints at the same time. While the women were consistently slower, given the need to accommodate the member with limited vision, each group did manage to arrive at the goal waterhole within the specified time frame on Day 5 to prepare signal fires. They *thought* they were going to be picked up and enthusiastically competed to provide the best rescue signal. They had SOS spelled out in rocks and in sand, smoky signal fires, billowing space blankets (one was provided per team) and lots of jumping and shouting when the 'rescue plane' flew over at 4 pm with myself and the survival instructors aboard.

However, we had other plans in store for the hot, tired and hungry teams. Out of the plane we dropped further instructions to the teams. Each was directed by different routes to conduct a night hike of over 30 km (18.6 miles) to a pick up point by 9 am the following morning. The men were further instructed that one of their team members (to be picked by a draw of straws) had been 'injured' from a campfire explosion and was required to bandage his eyes and be led by fellow team members over the night hike. This was to even the playing field between the men and women for this last sprint. A bright 9 am the following day found both teams gathered just short of the finish line where the camera crews, production team, local ranchers and yours truly waited to celebrate their survival experience.

What did we learn? There were definitely gender differences in how the groups coped with getting lost (a frequent experience for the less skilled women's team), frustration with failure to find plentiful food (sometimes, the fish in the waterholes just weren't biting), dealing with cold nights and blistering days, or stomaching the taste of chemical disinfectants for the water or laboriously boiling your water each evening and pouring it through a piece of cloth to filter out the particulate matter. Decision making was at times faulty (the men failed to notice one of their team members approaching sun stress) or inabilities to make decisions (the women frequently struggled to decide who was going to decide). Then there was that annoying camera person constantly documenting each grip, complaint and woe or the team member that walks/talks/eats too slow/fast. In extreme environments, little things become big things.

Each night, I reviewed the raw video footage sitting in my lorry in the middle of the Australian desert with the brilliance of the unpolluted night sky above me; the sounds of the desert wrapped around me and watched the teams morph from a collection of individuals with separate world views into a cohesive entity with a unified goal – make it to the finish. As fatigue took its toll, there was less drama and more conservation of energy on just getting to the next check point, i.e., getting the job done. Both teams got better. The men became more nurturing and paid closer attention to how everyone was doing; the women became more organized and task focused. In the end, they crossed the finish line together. None of them were what they were when they left but something more.

As a researcher, I was elated with the opportunity to be physically present as my teams met and overcame their challenges. As I debriefed each person, their insights into their own foibles and strengths were vastly illuminating. As I analyzed the data over the coming months and shared the results with the team members and the world, my understanding of the dynamic and complementary strengths of both genders grew just a tiny bit. As with any advance in any particular field, the contribution may have been minute compared to what we don't know but science is a process of accumulating minutia until sufficient thresholds are achieved for those great leaps forward.

It was a lesson that was also reflected in a more controlled comparison simulation study in 2005 at the Mars Society's Mars Desert Research Station (MDRS) in Utah (Bishop, Sundaresan, Pacos, Patricio, & Annes, 2005). Two teams, one male and one female, completed two week simulations of a Mars mission in a remote desert habitat. Both groups ran similar experiments over the two week mission, in the same environment; thus, the comparison was far more standardized than the teams in Australia.

As with the Australian teams, each group resolved similar situations with almost stereotypically gendered approaches (i.e., women focused on interpersonal relationships while men focused on task performance), yet they all achieved similar outcomes in the end. Coupled with the albeit small body of evidence on mixed gendered teams, there seems to be a slight edge for teams with mitigating strengths and weaknesses which has been reinforced by what I have seen of the shortcomings in same-gendered teams. The real question, the elephant in the room so to speak, is whether the feared problems surrounding sexual attraction and competition are outweighed by the benefits of having both genders present.

Is the concern about sexual complications legitimate? The answer, again, lies in the details. The now infamous incident involving uninvited sexual advances during the long duration simulation mission (Simulation of Flight of International Crew on Space Station -SFICSS) sponsored by the Russian Institute of Biomedical Problems certainly exemplified the worst of fears (Oberg, 2000; Sandal, 2001). A New Year's unwanted kiss led to a physical altercation between two male members, a complete interruption of the mission, and the subsequent withdrawal by the Japanese team member mid-term. Yet most agree that it was not the initial incident that was the culprit but the dismissive handling of the incident by the Russian mission control and then subsequent exacerbation by the press. Subsequent evidence (Rosenet, Jurion, Cazes & Bachelard, 2004) from Antarctic suggests that sexual attraction problems are most likely when young males and young females are present but not with any other configuration of age/gender combinations. Thus, the effective moderating principle for addressing sexual problems may simply be maturity.

The dynamics between a team and external support groups further impact performance negatively or positively (Kelly & Kanas, 1993, Kanas, Salnitskiy, Grund et al., 2000). While the communication logistics of long duration missions (e.g., 20 minute one-way lags to Mars) effectively rule out a reliance on real-time interaction with an Earthside Mission Control, the persistent evidence for the ease with which the "us versus them" mentality emerges between teams at remote locations and their support resources "back home" is undeniable. In almost every team I have investigated that relied on a mission support structure, a disconnect occurs at some point between the viewpoint of the team and the viewpoint of the mission controllers. It is not usually mission threatening, but that may be as much a matter of fortuitous happenstance that an emergency failed to occur that would be exacerbated by the communication difficulty. The problem is so pervasive that NASA has made efforts to provide interpersonal training of its mission control team as well as the astronaut crew. Differences in culture, perspective, professional roles, priorities, and situational factors tend to complicate shared perspective taking on both sides. This is true for within-team disconnects as well, which may represent even more acutely compromising effects. Isolation and shunning of a team member in a small group can quickly become a critical mental health issue and mission-threatening situation.

The need to keep team members integrated into the group was effectively demonstrated in my 2001 study of Norwegian Antarctic winter-overers at Troll Station (Bishop & Primeau, 2002). As the Antarctic winter progressed, one team member became distressed over news from home and withdrew from the group, ceasing to eat with them or maintain hygiene, essentially isolating himself. The group leader recognized the situation and sought advice via email as to how to counter the deteriorating condition. This concerned me greatly since there is a great sense of helplessness that is experienced by members in remote locations when distressing events are happening back home. The inability to be effectively present to make a difference in the outcome represents a significant psychological stressor. We settled on an approach that required attention to hygiene and attendance at meals reinforced by the leader (who was also the member's best friend) as well as the other two group members. Refusal to allow self-isolation and an insistence on maintaining attention to self-care were effective countermeasures to the situation and conditions vastly improved with resumption of a well-integrated team.

When I finally made it to Norway a year later to do a more extensive debrief of the team members, an unexpected but not terribly surprising (upon hindsight) finding emerged. The team doctor was much older than the other members (he in his 60's and they in their 20's) and well-established in his career.

As is frequently the case with individuals who must be away from their families for prolonged periods, time back home is like time frozen. In remote isolated environments, this is even truer because the consistency of the remote environment gets translated psychologically to the home situation as well. Yet, home environments are not stable and unchanging. Just the opposite! Their dynamic and ever-changing immersion in the larger world is also compounded by the requirement of family members to take on roles that they may have never had to enact due to the absence of the team member; from managing household expenses, to deciding what to put into the garden that year. Upon return of the absent member, not only has the family continued to grow and evolve but they are now very comfortable in their new roles and many times not at all interested in going back to the old status quo. Therefore, not only is the family *not* the same as expected but the member is no longer even *needed* in the same capacity as he/she was before leaving!

This reintegration challenge has long been recognized by the world's militaries who routinely remove members from their families as well as historical occupations that did the same, e.g., sailors. It is a recognized problem for astronauts and their families as well. Thus, when the physician's wife related similar difficulties this well-established couple of many decades had to deal with upon her husband's return, I was poignantly reminded that extreme environments affected not only those who occupied them but those who were left behind as well.

So, the challenge for those of us worrying about the human part of the equation has been how to choose the best fit individual to work within the best functioning group in an environment in which we must extrapolate from a very small sample with very limited experience.

Needless to say, arriving at answers has been a long path of discovery. The evidence for the contribution of inter-personal and intra-personal factors is substantial but fragmented and piecemeal. Psychological and psychosocial factors have been repeatedly acknowledged to significantly impact human behavior and performance in most challenging environments, especially those characterized by isolation and confinement, from the early reports by the National Commission on Space (1986) and the Space Science Board (1987) to comprehensive reviews of the field to date (Bishop, 2010, Kanas et al., 2009). We have gathered a multitude of informational bits across myriad environments and mission types. There has been a gradual but reluctant recognition of the importance of psychological and sociocultural factors for group functioning and their role as critical components to mission success. But progress has been slow and frustrating.

The factors that have yet to be adequately screened and validated are still numerous. As I and others have reiterated repeatedly over the years, the discovery process has been fraught with multiple challenges. There is little funding to support the intensive endeavors needed to rigorously test various factors known to impact team and individual performance (e.g., crew composition variables, group interaction variables, interpersonal skills, leadership factors, individual hardiness, response and coping with stress), analog expeditions are scattered worldwide and large scale natural or simulated expeditions virtually nonexistent. Research on actual space crews is further compromised by incomplete participation by crew members, multicultural issues, and at times only the barest definition of a "group," e.g., two-person ISS expeditions. Researchers themselves are spread across various disciplines, which impairs knowledge sharing and growth (Bishop, 2004).

The opportunistic nature of studying real-world teams in extreme environments as our best answer to full fidelity comparable scenarios also provides significant challenges. There is minimal input from researchers regarding team composition, structure, leadership, expedition schedules, and so forth. It is usually incumbent on the researcher to find a way to interleave their research needs among the operationally driven parameters. Sample sizes are typically very small and participation in research protocols strictly voluntary, contributing to high rates of drop-outs and missing data, especially when the expeditions are having difficulties.

So, one might ask, is the research on real-world groups valid for future space crews? This is the paradox faced by researchers tasked with identifying the key selection factors for future long duration space crews. In laboratory studies, the very attributes of the environment that have the greatest impact on performance are removed (e.g., real danger, uncontrolled events, situational ambiguity, uncertainty, the interaction with the extreme environment itself). Complexity is a key defining trait of stressed operational environments. Reliance on laboratory studies and the presumption of broad generalizability particularly for research on high-stress, high-risk environments, is highly likely to lead to dissociation between actual operational findings and laboratory and experimental studies. Yet there simply is NO existing available "group" of space-based individuals to investigate. Thus, terrestrial analogs are our only viable alternative. The challenge has been to marry the needed control and rigor that would be inherent in the highly controlled and monitored space environment with terrestrial environments that present a significant component of risk and stress that would be "comparable" with conditions in a space environment.

So the journey of discovery continues. The adventure is ongoing. We've still more questions than answers but we know a bit more than we did yesterday. Patterns and commonalities are emerging across environments. We have a better chance of selecting "good fits" today than we did yesterday. Tomorrow, perhaps, we'll be able to reach the vaunted "best fit" goal.

Until then, Ad Astra!

References

1. Bishop SL, Santy PA & Faulk D. When Teams Fail: A Case Study of the Huautla Deep Caving Expedition, *Society of Automotive Engineers.* 29th International Conference on Environmental Systems (ICES),1999.

2. Bishop, SL. A Comparison of Male and Female Teams in Surviving the Australian Outback, *Proceedings from the Human Systems Conference*, Nassau Bay, Texas, June 20-22, 2001.
3. Bishop SL, Grobler LC & SchjØll O. Relationship of Psychological and Physiological Parameters during a 350 km Arctic Ski Expedition: A Case Study, *Acta Astronautica*, Vol. 49, No. 3-10, pp. 261-270, 2001.
4. Bishop S. Evaluating Teams in Extreme Environments: Deep Caving, Polar and Desert Expeditions, *Proceedings from the 32nd International Conference on Environmental Systems (ICES)*, San Antonio, Texas, USA, July 2002.
5. Bishop S & Primeau L. Through the Long Night: Stress and Group Dynamics in Antarctica, *Proceeding from the 53rd International Astronautical Congress (IAC) of the International Astronautical Federation (IAF), International Academy of Astronautics (IAA), and International Institute of Space Law (IISL)*, Human Factors for Long Duration Spaceflight, World Space Congress, October 10-19, Houston, Texas, 2002.
6. Bishop S. Evaluating Teams in Extreme Environments: From Issues to Answers, International Space Life Sciences Working Group (ISLSWG) Group Interactions Workshop, *Aviation, Space and Environmental Medicine,* July, Vol. 75, no. supplement 1, pp. C14-C21(1), 2004.
7. Bishop SL, Sundaresan A, Pacros A, Patricio R, & Annes R. A Comparison of Homogeneous Male and Female Teams in a Mars Simulation, *Proceedings from the 56th International Astronautical Congress*, Fukuoka, Japan, October 17-21, 2005.
8. Bishop SL. Moving to Mars: There and Back Again. Stress and the Psychology and Culture of Crew and Astronaut, *Journal of Cosmology*, V12, pp. 3711-3722, 2010.
9. Freedman JL. Crowding and Behavior. 1975, New York: Viking.
10. Fortney SM, Beckett WS, Carpenter AJ, Davis J, Drew H, LaFrance ND, Rock JA, Tankersley C. & Vroman NB. Changes in plasma volume during bed rest: effects of menstrual cycle and estrogen administration. *The American Physiological Society,* pp. 525-533, 1988.
11. Kanas N, Salnitskiy V, Grund EM, Gushin V, Weiss DS, Kozerenko O, Sled A, & Marmar CR. Interpersonal and cultural issues involving crews

and ground personnel during Shuttle/Mir space missions. *Aviation, Space Environmental Medicine*, 71(9, Suppl.): A11-6, 2000.

12. Kanas N, Sandal G, Boyd JE, Gushin VI, Manzey D, North R, Leon GR, Suedfeld P, Bishop S, Fiedler ER, Inoue N, Johannes B, Kealey DJ, Kraft N, Matsuzaki I, Musson D, Palinkas LA, Salnitskiy VP, Sipes W, Stuster J, & Wang J. Psychology and culture during long-duration space missions. *Acta Astronautica,* 64: 659–677, 2009.

13. Kelly AD & Kanas N. Communication between space crews and ground personnel: A survey of astronauts and cosmonauts. *Aviation, Space, and Environmental Medicine*, 64: 795-800, 1993.

14. National Commission on Space, Pioneering the Space Frontier. New York: Bantam Books, 1986.

15. Oberg J, 2000. http://www.jamesoberg.com/04142000assualt_rus.html, accessed on 1/10/14.

16. Palinkas LA. Effects of physical and social environments on the health and well-being of Antarctic winter-over personnel. *Environment and Behavior,* 23, 782-799, 1991.

17. Robinson, KS. The Martians, Spectra/Bantam Dell/Random House, 1999.

18. Rosenet E, Jurion S, Cazes G & Bachelard C. Mixed-gender groups: coping strategies and factors of psychological adaptation in a polar environment. *Aviation Space and Environmental Medicine,* 75(7 Suppl): C10-3, 2004.

19. Sandal GM. Crew tension during a space station simulation. *Environment and Behavior*, Jan; 33(1):134-50, 2001.

20. Santy PA, Holland AW, Looper L & Marcondes-North R. Multicultural Factors in the Space Environment: Results of an International Shuttle Crew Debrief. *Aviation, Space and Environmental Medicine,* 64:3, 196-200, 1993.

21. Space Science Board. A strategy for space biology and medical science, Washington, D.C: National Academy Press, 1987.

22. Steele, CW. Huautla: Thirty Years in One of the World's Deepest Caves, Cave Books, 2009

Empirical, Clinical and First Person Neuroscience: Wanting it All

Jonathan Cole

Jonathan Cole is a consultant in Clinical Neurophysiology at Poole Hospital and professor at the University of Bournemouth. After medical school in Oxford and The Middlesex Hospital, he undertook research in Oxford before training in clinical neurophysiology in Southampton. His research has been in sensory loss and its consequences for motor control, spinal cord injury, pain and tremor. He is a Past president of the British Society for Clinical Neurophysiology, was chair of the 2006 International Congress, and is current secretary of the European Chapter in Clinical Neurophysiology. In addition to empirical research he also has an interest in the subjective experience of neurological impairment and studied with Oliver Sacks. His books include Pride and Daily Marathon, *on Ian Waterman who lives without proprioception,* Still Lives *(on spinal cord injury) and* The Invisible Smile *(on living without facial expression). His second biography of Ian,* Losing Touch *is in press with OUP. He has also collaborated with artists in both theatre (*The Man Who, The Articulate Hand*) and dance, with the choreographer Siobhan Davies.*

Apart from the more usual daydreams of adolescence, focussing on unattainable girls whose names I could still recite, I remember wondering what it would be like to be conscious but completely disconnected to one's sensory and motor apparatus; being able to see and hear, but not move or feel from the body, how would that affect, for instance, my perception of space and distance? Thus my initial interest in neuroscience began at school with a brain-in-a-vat thought experiment. At that time I was fascinated by biology and particularly by physiology, how animals work, as a way to understand how we work. It was not that there was any experiment or experience that was a eureka moment; I just wanted to know what went on inside. But as my interests developed I was also reading literature since that also seemed to allow one to understand our internal lives albeit from a very different perspective and level. In England at that time we had to choose between studying arts/humanities and science from 16 onwards, and since the pull towards physiology was greater, science won.

When it came to university, left to my own devices, I would have continued with physiology and especially neurophysiology. [At that time there were few courses in neurophysiology alone and the term 'neuroscience,' if it had been invented, was not in common currency.] My headmaster had other ideas, however, and guided me towards medicine. This was sold as both a profession but also as a passport to many more differing areas than physiology alone. One other event occurred, which was a light-bulb memory. My biology teacher invited me into his hallowed office, sat me down, gave me a cup of tea and asked if I had thought of applying to Oxford or Cambridge. I had not until then, but afterwards wandered around in a daze contemplating such a thing. I grew up in the Midlands, in Birmingham, which had good train links to Oxford but none to Cambridge. A day trip to look round the dreaming spires, and that was it.

At Oxford I was very fortunate to be taught by some outstanding neurophysiologists in David Whitteridge, Charles Phillips and George Gordon. Not only inspiring, they were also interested in me too. I was late for one of Prof Phillips' lectures one morning and turned up soaked, with long dank student hair everywhere. He was waiting outside, so I apologised. 'No problem, Jonathan,' he said, 'I could not start without you.' I was astonished that he knew my name. Neurophysiology fascinated me and after my first degree I stayed on to research, with my college tutor George Gordon becoming my supervisor, on how the cerebral cortex gates its own sensory inputs. Golden days, with stretching of my intellect and my horizons in a community where the currency was ideas.

My parents were neither young nor wealthy and I thought I owed it to them to complete my medical studies. I loved Oxford but I also had the nagging thought that I should sample the 'real' world, and so went to The Middlesex Hospital, London. It was fine but medical studies were not; the long lists of facts and little intellectual curiosity were really difficult; no one called me by my name, few even knew it. Though I qualified and made some good friends, memories are grey and sparse. Then 3 years as a hospital doctor and passing of the exit exam, for Membership of the Royal College of Physicians, before I returned, thanks to George, to Oxford and more academic research in sensorimotor neurophysiology. After that I felt I wanted to try to employ my research skills in a clinical setting so I moved to Southampton for a project on pain in spinal cord injury. This was followed by training in Clinical Neurophysiology and a lifetime trying to combine clinical work with empirical research. There was, however, one other matter.

As a hard up student in London buying new hardback books was an infrequent luxury, but reviews led me to one book which has influenced me profoundly. *Awakenings* seemed to fuse medicine and literature, dispassionate recording of patients' experiences with empathetic understanding in a way touched by genius. I contacted Oliver Sacks and asked to study with him. He very generously agreed and so, for two months in the mid 1970's, I went to New York and shadowed him, feeling intellectually out of my depth, soaking up neurology, psychology, philosophy, music, literature, gloriously slummy Bronx diners, posh Fifth Avenue restaurants and much more besides. Oliver showed me the possibilities of exploring and writing about what it was like to live with chronic neurological conditions, and the importance of listening to people tell their stories.

Whilst there a British film crew was out there – his fame was spreading – for a programme on a subject with Tourette's. I remember spending time with the crew trying to unblock Oliver's loo – an early glimpse of the magic of filming - and hanging out with the subject. I learnt how much you can learn from just sitting with someone, without a white coat, or a microphone and camera; he started talking about aspects of living with Tourette's that he never disclosed on camera.

So, I was equipped with an intellectual fascination with neuroscience, and a desire to understand others' situations and explore it in its necessary depth. There was one further impetus which is more difficult to explain and which has been my main day job; clinical medicine. The simple relationship between doctor and patient is not easy to discuss, personal and individual though it is. But the satisfaction of having put a nervous patient at ease and tried one's best to help them is a continual source of satisfaction shared with all members of the health community. Though not often discussed it is still a profound pleasure and motivation.

The project in Southampton also allowed time for other research and I developed a novel technique for looking at motor control and tremor. Through that I met a man, Ian Waterman, whose condition immediately fascinated me, in the mid-1980s, and still does to this day. Twelve years before we met he had lost, below the neck, the sensations of touch and of movement and position sense or proprioception, though with retained perception of pain and temperature and with normal motor nerve function. When it happened he was completely unable to move, since the brain had no feedback of where the body and limbs were in space. Ian had relearnt to move by mental concentration and visual supervision in an astonishing way. The condition, *sensory neuronopathy syndrome*, remains very rare and his functional recovery unique.

I set out to explore his condition through a series of experiments focussing on his perceptions of force and effort without peripheral feedback.

His perception of a weight held in the hand is the same as control subjects (he could distinguish weights which differed by 5g in 100g given one after the other) if he can see the weight and move it up and down; he makes a set movement and sees whether it moves faster or further than the last weight. But with eyes shut he can only judge weights differing by 100%. Quite how he does this is still unknown but it could well be using peripheral receptors and nerves that still exist and which normally relay feelings of tension or fatigue.

In a further experiment we looked this time at Ian's sense of effort. The task was to match forces between hands in a grip test, before and after the muscles had been fatigued so much that the forces they could produce were only 50% of a rested muscle. Under these circumstances control subjects think their fatigued muscle matches the other when it actually is producing 40% less force, because receptor desensitisation after fatigue reduces spindle and tendon organ reafference so that weights feel lighter than on the unfatigued side. In contrast Ian perceives a match when putting 100% more effort in; since the muscle is producing 50% less force he doubles the central sense of effort his puts into the task. In this way we showed that in controls there is a complex relationship between peripheral fatigue, central effort, and these two origins of the perception of force and effort.

One question was whether Ian is aware of the movements he makes. We looked at this using two methods. In the first he had to press down with his index finger on a small torque force and match it with the other index finger. Without feedback to learn Ian could not do this, either matching one side to the other, or doing the two together. We suggested he did not have knowledge of motor command, i.e. not access to any corollary discharge. We also used a technique, transcranial magnetic stimulation, to excite his motor cortex to produce a small twitch movement of the opposite arm. Then we asked Ian if he could judge when he moved, as a result of (for him) the non-intentional discharge of his motor output. The answer was "no." Of course he has knowledge of what he does, since he has to consciously intend to any action, but this intention is towards what he intends to do, what he asks the motor apparatus to do, and not a lower level awareness of what the motor cortex has commanded.

Ian has recovered to produce more complex movements than some others with the condition. One reason for this is his ability, built up over years, to run forward motor programmes in the brain to allow movements to unfurl. One way to show this is through a mirror drawing task. Subject GL, similar to Ian, had done this with the Quebec group some years ago. Asked to draw round a simple Star of David on a desk with her finger with visual feedback mirror reversed, she was able to do this easily the first time whereas control subjects were stuck at the points of the star, when you have to go the opposite way to where you see yourself going. They concluded, reasonable enough, that GL could do mirror drawing because she did not have any mismatch between where she *saw* herself going and where she *felt* herself going through proprioception.

Many years later, with Chris Miall in Birmingham, we used a virtual mirror, a graphics tablet and more complex patterns which subjects had to trace round using mirror reversed feedback. Ian was stuck at the corners, like control subjects. Our conclusion was that Ian, like controls, is stuck not because of a mismatch between vision of the arm moving and a feeling of where the arm is going from proprioception, but because the mismatch is between visual feedback of where the hands goes and an internally generated motor programme of where the hand is planned to go next. The problem at the points of the star is felt as conflict between where you *see* the pointer going and where you *will* it to go, rather than between where it is seen and where it is felt. The experiment was relatively simple but important since it showed that Ian uses these predictive forward motor programmes, whereas GL used on line visual feedback, which was usually slower and less efficient, but in this particular task allowed her to outperform controls.

Ian does not like doctors but agreed to take part in work which now has extended over 3 decades because I was not solely interested in empirical approaches. I immediately started asking what it was like for him to live like this; the first time a doctor had gone beyond the clinical, third person approach to ask about his first person subjective experience. As a result we collaborated on Ian's biography, *Pride and a Daily Marathon* (1991, 1995), on a BBC TV Horizon documentary, *The Man Who Lost His Body* (http://tinyurl.com/manlostbody), (1998), and to Ian being portrayed by Peter Brook in his play based *L'Homme Qui*, (1994) based on Sacks' *The Man Who Mistook His Wife For a Hat*. More recently Ian's vignette from that has been extended and deepened for Brook's work, *The Valley of Astonishment* (2014). We have also prepared for publication a second volume of Ian's biography, covering his life over the last 25 years, and what it is like to do research and be the subject of media and theatrical portrayals. It also gives Ian's take on being a neuroscience subject *(in press, OUP).* In this Ian explains why he agreed to my interest,

> 'What made the haul so difficult was that I did not have the vocabulary to explain it. I had spent time talking to doctors but they did not know what it meant, to me. The first doctor to ask how I live with it was Jonathan. He asked about the human side as well as the clinical; that was revelation in those days. It is crucial to

> understand what it is like, and Jonathan was the first person to actually listen.'

Without my interest in Ian as a person it is doubtful he would have consented to so many experiments. For me the two approaches are parts of the

I was giving a lecture in Vancouver on work with Ian when afterwards I was approached by someone from The MIT Press. Would I like to write a book? Though many in neuroscience, especially in their younger period, hesitate to give so much effort towards something which is not counted in research assessment exercises, I leapt at the chance, all the more since there was a subject which had interested me for some time. I knew it needed sustained and deep exploration and that a book was the ideal way to approach it.

The book, *About Face*, was an exploration of the relations between self and face. My approach might be considered perverse. The world is awash with famous and beautiful faces; I wanted the experience of those for whom the face is a problem. What of those who are blind and who cannot use the faces of others? For one man, blind from birth, everything was in the voice. For John Hull who lost his sight in middle age, because face represented others to him, through visual memory, blindness led to a loss of the other which drove him into isolation and depression, until he learnt to replace vision with hearing.

> 'The horror of being faceless, of forgetting one's own appearance, of having no face. The face is the mirror image of the soul.' Hull, 1992.

I talked with people who live with autism, who cannot comprehend all the movements the face does, and to those with Möbius Syndrome who have a congenital absence of facial expression, and to those with facial disfigurement. One of them, James Partridge, is the chief executive on a UK charity providing social support for those with disfigurement (a term they use reluctantly) as well as education for health care workers and others. Subsequently James invited me to sit on the scientific advisory panel of the charity, which I have felt privileged to do since.

About Face built up the results of doubt in relation to the face, and concluded with its importance for interpersonal relatedness. More controversially, perhaps, I suggested that a mobile expressive face allowed the development of the more complex emotional states which characterise our species. The fact, being so obvious and so given, is not always given sufficient import in consideration of who and what we are. *About Face* tried to show this by asking about life without it. I felt I was exploring new ground and during the time of researching and writing it I mixed exhilaration with dread, since I sought no one else's advice, and was worried that it would be criticised for some gaping mistake or hole. Fortunately no one has, (yet).

I had spent two years working with people with spinal cord injury. Disappointingly I showed the treatment (spinal cord stimulation) was ineffective for chronic pain, though I did suggest why, on physiological principles. I also was very aware that, like most clinical workers, I had shied away from the big question of how people live with the condition. Try to imagine, in a second, going from being whole to being unable to move or feel below the neck, forever, and to being doubly incontinent. Then add in chronic pain, which many with spinal cord injury patients also have. A larger loss of embodiment and a bigger threat to one's self is difficult to imagine. Health care workers tended not to ask, fearful of the answer. This no-go area was described by Reverend Albert Bull, a paraplegic Army chaplain who had spent eighteen painful months at various hospitals before arriving at Stoke Mandeville in 1944, when he wrote, 'The first duty of a paraplegic is to cheer up his visitors.'

In *Still Lives* I explored the lives of twelve people with tetraplegia. This was the same story twelve times, but because people are all so different, their stories and experiences differed too. The book moved from two people whose experience was poor, and who might reinforce the reader's expectations, to slowly overturn this and reveal something of the extraordinary rich and varied lives that many – most? - people with tetraplegia live. One man described it thus,

> 'You just cannot substitute for the experience of being able to use this wonderful piece of equipment, the body, be it running, riding or shagging. My greatest passion was horse riding. The sheer enjoyment and freedom of being able to go hell for leather across the forest on the back of this living being with communication and some measure of control, but not too much. It was awe-inspiring and wonderful, a whole body experience with sheer total involvement and the physical contact. People say why not go to 'Riding for the Disabled'. I don't want to sit on a horse and be led round a fucking paddock. That's nothing compared to what I was doing.'

In contrast many became adapted;

> 'I still view my body as whole; it's just motionless. I'm not a head on a bag of potatoes. I still know it's there, I still like it; I like to see it, it is still me, and I am still it, totally.'

And amazingly for some their new embodied existence becomes their new normal, very like their old normal,

> 'Now I can almost kid myself that I can feel something when I sit in a chair, even though I know I cannot. It feels exactly the same sitting in a chair now to before I was injured. It can't but it does. My mind makes me think I am like you over there. It learns what is the norm for this body.'

The need to feel something in the body was made explicit by someone who lived with severe pain in their body. [If the brain cannot feel from the body then pain like phantom limb pain results.]

> 'My physical pain is in the hands and down the legs and in the feet, the whole time. It is hot, on the inside, like needles. The feet feel as though someone has a bicycle pump on them, they feel as though they are about to explode… Nothing makes it worse or better.

> I cannot do anything myself to get a connection with my body. If I pinch my legs it is numb, so having the pain puts me in touch with my body. The pain is the connection - my friend the pain. It is almost comfortable, almost my friend.'

For some people with tetraplegia it is no disaster but just another part of their lives. Mike Oliver, a distinguished academic in the disability movement and himself tetraplegic since a young man told me,

> 'Breaking my neck broke that mould and gave me an alternative possibility. It changed the possibility of who I could become. 40 years later I am a professor of disability studies, I have one marriage behind me and I am happily married again, I have grandchildren and have been all over the world. I have had a good life. It was the best thing that had even happened to me. I was a working class yobbo with a failed education, not very good at relationships in a job that I did not like and I probably would have gone on to drink and smoke too much… I was a promising sportsman but had failed in that too. If I had not broken my neck I would not be a professor in a university.'

I asked another man whether, if I had a magic wand to make him whole again, would he take it,

> 'Why might I not want to? Well, I think 30 years down the line I am used to being who I am. I would actually have to go through a grieving process for losing me as I am now. That would be very strange. It might come across as negating all the love and friendship as a result of being spinally injured. Because I am as happy as I am I don't really feel the need to want to change.

I returned to explore the face when a received a phone call from a woman, Henrietta Spalding, who had Möbius Syndrome. She had read *About Face* and thought it described her condition inadequately. Rather than argue I suggested we collaborate on a book purely on that condition.

The experiences of those with immobile inexpressive faces from birth had so much to teach us all about what the face does. She agreed and I would write and then let her judge the tone and truth of what I had done from talking with a number of people with the condition. In *About Face* one man with Möbius, in middle age and a successful priest had written of a sort of disconnection between what he felt and thought.

> 'I have a notion which has stayed with me over much of my life - that it is possible to live in your head, entirely in my head. Whether that came out of my facial problem I don't know. When meeting my wife, I think initially I was thinking I was in love with her. It was some time later when I realised that I really felt in love. I think I get trapped in my mind. I sort of think happy or I think sad, not really saying or recognising actually feeling happy or feeling sad. Perhaps I have had a difficulty in recognising that which I'm putting a name to is not a thought at all but it is a feeling, maybe I have to intellectualise mood. I have to say this thought is a happy thought and therefore I am happy.'

Some people we met subsequently have the same experience. One woman, who I called Celia, remembered her childhood and all the operations she had,

> 'I did not do ballet or horse riding; I did hospitals and operations. I had the eye doctor and the foot doctor and a speech therapist, and a face doctor. My limitations were a fact of life. Not being able to see the blackboard, or not being able to see someone over there. [People with Möbius can also have eye and other problems too.] I never thought I was a person; I used to think I was a collection of bits. I thought I had all these different doctors to look after all the different bits. 'Celia' was not there; that was a name people called the collection of bits. I did not like my feet; I liked my spirit, I liked my brain; I knew I had a brain. Even though I was a collection of bits I always knew there was something strong inside

> that I had a mental dialogue with, but it was not the physical body; it was very separate from the physical.'

As a child she remembers little inner emotional life. She developed this as a young adult, at university and then teaching English in Spain, when she developed, consciously, gesture and expressive speech and, for the first time, began to experience the emotions behind them.

> 'All my gesture is voluntary, even now aged 40. Everything I do, I think about… All the things I am doing, whether turning my head or moving my hands, is self-taught. As a child and teenager, I had the words, but words are not appropriate. When I was a child I could not gesture, because I was a collection of bits. My body was not me, so expression in it, with it, would not be from me either. It was not a joined up feeling. There was huge bit missing; with the lack of balance, mobility, and problems with coordination, you don't get a sense of self… I could see everything and wanted to communicate but I could not *do* anything. It makes you so different.
>
> 'With Möbius you have to be much more wordy and articulate and this requires intelligence and so it can be hard and tiring. For me the word is stronger than facial expression. Without the word, how to express the feeling? I am interested now in non-facial aspects; gesture and tone of voice. Gesture is part of language, is a language and people with Möbius do not always learn it; they must be taught. Without gesture, thought is impoverished, as is language without gesture and thought without language… Use the hands and you feel it too.
>
> 'As a child I did not know I was missing out. To reach out to others I needed embodied expression – feeling. As you grow up the social feedback from others has far more meaning than as a child. A meaningful smile from you triggers an emotional response from me. As a teenager I was articulate but this was not sufficient. When I express being happy that has to be vocal and intellectual… There is an element of artificiality in the expression but not the feeling.

> Even with emotion now I have two sorts of happy, intellectual happy to express, and happy, happy. I can be happy, not think happy, but to express I have to think.'

My professional interests have been in clinical medicine and empirical research, but also in understanding how people with chronic impairment live with their conditions. One reason for this, apart from my interest honed by time with Oliver Sacks, is contained in the following;

> 'Science manipulates things and gives up living in them.' -Maurice Merleau-Ponty, 1964.

Science is enormously powerful but there are insights available only from those individuals with the condition which teach us not only a huge amount about their condition but also about how we all live.

This short intellectual journey has focussed on publications and writing and not on my day job, as a doctor, though I did acknowledge this earlier. Of equal or more importance has been my wife, Sue, of over 30 years and our four well-adjusted, life-enhancing daughters. Watching them grow and become independent has been a deep enduring joy, and it remains just as much fun watching them bicker and josh each other when they get together. When they were small I used to write and direct a children's play each year in our village, clocking up round a dozen from Ben Hur, Macbeth, (the Scottish Panto), to Scott of the Antarctic. And for 50 years or more I have run and cycled at serious - though declining - distances and speeds. Medicine, science, writing, family, sport and more; as Ian Waterman says, 'Life is no bloody rehearsal.' Seize it by the throat.

References

Cole J D and Sedgwick E M - The perceptions of force and of movement in a man without large myelinated sensory afferents below the neck. J. Physiol. 449, (1992), 503-515.
Fleury M, Bard C, Teasdale N, Paillard J, Cole J D, Lajoie Y and Lamarre Y – Weight judgement: the discrimination capacity of a deafferented subject. Brain, 1995, 118, 1149-1156.
Cole J. Pride and a Daily Marathon. Cambridge, MA and London: The MIT Press, 1995.
Cole J. About Face. Cambridge, MA and London: The MIT Press, 1998.
Cole J. Still Lives. Cambridge, MA and London: The MIT Press, 2004.
Cole J and Spalding H, 2008. The Invisible Smile, Oxford; OUP
Hull, J. Touching the Rock; an experience of blindness. New York: Random House.
Luu BL Day BL, Cole JD, Fitzpatrick RC. The fusimotor and reafferent origin of the sense of force and weight. J. Physiol, 2011, 589:3135-47.
Maurice Merleau-Ponty, 1964. From 'Eye and Mind,' in The Primacy of Perception, Chicago: Northwestern University Press, p159.
Miall RC and Cole JD. Evidence for stronger visuo-motor than visuo-proprioceptive conflict during mirror drawing performed by a deafferented subject and control subjects. Exp Brain Res. 2007; 176(3): 432-9.

The Cold Hard Face of Discovery in Remote Antarctica

John A. Long

John Long is currently Strategic professor in Palaeontology at Flinders University, prior to that he was Vice President at the Los Angeles County Museum of Natural History (2009-2012), Head of Sciences at Museum Victoria (2004-2009) and Curator at the WA Museum (1989-2004). He researches the early evolution of vertebrates, and his discoveries includes the oldest evidence for sex in vertebrates, published this year in Nature. He has authored over 200 scientific papers and popular science articles and some 28 books. His awards include the 2001 Eureka Prize for the Public Promotion of Science, the 2008 Australasian Science Prize, the 2011 Royal Society of Victoria Research Medal and the 2014 Verco Medal from the Royal Society of South Australia.

Throughout my life the act of discovery, finding things new to science, has happened serendipitously and unexpectedly, like experiencing the joy of snow falling on a sunny day. A memory of such days lingers in the back of my brain as I recall my two expeditions working in Antarctica back in the late 1980s and early 1990s. Some days it was positively balmy, with temperatures hovering at around 0°C, and the clear radiant polar sun belted down on our faces and reflected warmly under our chins from the surrounding snow fields. On days like this we could strip down to work in tee shirts, with our spirits buoyed up by the hope of a few continuous days of clear weather. We could muster ourselves at such times to cover great distances on foot to reach the most remote jagged outcrops in our quest for finding fossils.

The work was physically hard and demanding, but the pay-off is enormous when one makes a significant scientific discovery. It was the thought of making such discoveries in an unexplored area that first lead me down there, despite the known dangers, the isolation from home and the psychological traumas that deep field expeditioners usually have to face. We went sledging through about 400 miles of unexplored section of the Transantarctic Mountains, in Southern Victoria Land, charting unknown territory where no human had ever been before. Such feelings raise the adrenalin and sharpen the senses. The imminent threat of danger is ever present on such journeys, especially during the days we spent sledging between sites, traversing vast snow fields that lightly blanketed hidden crevasses. When climbing steep snow laden mountains we entered the deadly domain of unexpected avalanches, or risked sudden slips down steep icy slopes. But the fossils we found, the discoveries we made for science, pulled me through the arduous days. The fossils made it all worth the risk.

I recall clearly the days in late December 1991. We were working close to the polar plateau where the last vestiges of enormous mountain ranges poked their gnarly green and grey layered heads out of the nearly 2 km thick sea of snow and ice enclosing them. Our mission was to explore the the peaks to study the geology and paleontology, bringing back specimens of past life from the Devonian period that once inhabited the region about 380 million years ago. In those days there was no Antarctica, as it was subsumed within the vast supercontinent of Gondwana (http://tinyurl.com/triassicmap). Antarctica was then land locked by a fusion of undefined continental masses that today we identify as Australia, India, Africa and South America.

Back in the middle part of the Devonian there was no large animal life on land, only a small diversity of primitive plants that survived within a close range of the waterways. Here lived horsetails, ferns, mosses, lichens and fungi that formed low, sparse forests with a canopy only 10-20 feet high. It was a premonition to the great 100 foot high *Lepidodendron* and *Archaeopteris* forests that would cover the world in the next 30 million years as land plants rapidly evolved in response to increased surges of oxygen entering into the atmosphere towards the end of the Devonian. But in the Middle Devonian, here in Antarctica, the forests on land were only just getting into gear. Small spider-like creatures (trigonotarbids) and soft bodied worms may have inhabited the first soils that formed at this time, but we have no fossil record of them from

Antarctica at this stage. The rivers and lakes were home to a huge diversity of early fishes, at least 50 species have now been found due to our expeditions and fossils found on earlier expeditions in the 1970s that were collected at sites further to the north of where we were based.

The adventure began the day our sledges and toboggans, packed with everything we would need to survive the trip were hastily ejected out of the rumbling backside of a VXE6 Squadron C130 Hercules on the Darwin Glacier (http://tinyurl.com/darwinglacier) on grey day in early November.

After a quick test that our radio comms were operational, the plane took off from the glacier and disappeared into the distant skies. Then it hit me that we four humans were alone in possibly the most remote part of the planet, and had only our equipment, supplies and our training to ensure our survival until the plane would one day return, and scoop us up again, in about 2 months time.

The expedition set off the next day from the Darwin glacier landing site to reach the first outcrops about 30 miles away where we would set up the first base camp and begin our work in earnest. The four of us would travel using 2 motorised toboggans each pulling two Nansen sledges loaded with our fuel, food, field equipment and our scant personal gear. Each person had a job, either driving or being the sledge man at the back of the second sledge. The two sledges pulled by each toboggan had about 30 yards of rope between them, so that if the front of the sledging train went down a crevasse, the rear sledge man could quickly anchor the entire entourage to the ice and then rescue the toboggan driver who was always roped to the rear sledgeman.

On that first day I rode behind the back of the second sledge and carefully watched the driver ahead as we ventured into new territory, crossing a large snow field that could be riddled with crevasses. That day nothing dramatic happened, it was smooth traveling all day, but as we entered the mountainous region where we intended to camp the winds suddenly picked up and the temperature began plummeting. Before long we were desperate to finds a place to set up our two-man pyramid tents within some sort of shelter of the mountain, but the howling, biting winds kept lashing us at every place we tried to set up camp. In the end we four people wrestled each tent and finally got them set up in the relentless winds, allowing us eventually to crawl into the shelter of our tents, exhausted after the days sojourn.

As the expedition progressed we experienced many days of good weather and so traveling was pleasant, but there were also other days when storms fueled by the gravity-fed katabatic winds came rolling down the polar plateau out of nowhere, making us prisoners inside our tents for several days on end. Yet between the bouts of bad weather and days sledging to new sites we went out searching and collecting specimens.

Our finds included a set of fossil trackways resembling a pair of ancient motorcycle tire tracks, preserved in flat sheets of sandstone. These were made by an enormous sea scorpion (*eurypterid* - http://tinyurl.com/seascorpion), about 6 feet long that lumbered out of the primeval seas onto land around 400 million years ago, when global oxygen levels soared high enough to allow marine creatures to simply keep growing to unprecedented sizes.

Eventually our travels took us to mountains where the Aztec Siltstone outcropped, so named as the type section of exposed rock looked like an ancient Aztec Pyramid. This was a thick unit of sedimentary rocks characterised by many layers of differing colors: bands of red and green mudstone, black shales, but predominantly grey to yellow sandstones. It was in this unit we would focus on finding fish fossils. The distinctive Aztec layers were visible from afar, looking through binoculars across the wide glaciers, we could pick out the fossil rich layers from at least 20 miles away. The 380 million year old layers slashed through the mountains like a crazy Pollock painting and were sometimes crammed between thick black layers of intruded volcanic doleritic rock, or capped above by bands of striped Carboniferous coal measures.

We would sometimes find new fossil sites and give them nicknames. At one site the beds of rock were so packed full of fish fossils, we named the site ‘Fish Hotel’ as all the beds were full. The discoveries made at the site included many examples of armoured placoderm fish plates, jaws of giant predatory bony fishes called tristichopterids (some species were up to 10 feet long), and spines from an extinct group of shark like fishes called ‘acanthodians’. To me some of the most important discoveries we made on that trip were of small but highly significant things, such as the teeth of ancient sharks that once inhabited freshwater rivers.

I recall finding isolated teeth of sharks that were completely new to science at the Fish Hotel site, and scribbling down in my notebook that night that these represented, as far as I knew, the first great radiation of sharks on the Gondwana supercontinent. In addition to the other 2 species named from previous expeditions it meant that this one unit of sedimentary rock had at least 5 different varieties of early sharks living in its ancient rivers and lakes. The sites produced 3 new kinds of fossil sharks, one being at least 6 feet in length, making it the largest shark in the world at that time. I eventually named each of the three new shark species after my 3 colleagues that were my buddies on the expedition.

This discovery would have important ramifications later in life as in the mid-1990s on a fossil hunting expedition to the Cederberg Ranges in South Africa, I rediscovered these same species of shark for the first time on the African continent. It was a dismal cloudy day, lightly drizzling with rain. Baboons loitering on the cliffs above where I was digging were hectoring me by throwing small rocks at me and squealing loudly at my presence. Much later on a high mountain range in central Victoria, Australia, I found a tiny sharks tooth from a new fossil site, and it was one of the Antarctic fossils shark species. To me the real power of discovery comes not with the single find of something interesting, but is unlocked in the enlightenment of serial finds in different places. When they all join together in your head you can embrace a deeper meaning to each find as you can draw the imaginary boundaries and mark out the ancient geographical range of a species. Such finds help us prove ancient landmasses were indeed close together, supporting the idea of a conjoined Gondwana. These sharks lived in large freshwater rivers that joined up and met with similar rivers across the 3 continents, a sublime and satisfying proof to me that these land masses were once all joined together.

For most of that expedition in Antarctica things went well and we collected a large amount of fossils. But not all the time. Collecting fossils in Antarctica has its unexpected dangers. When one sets out to collect fossils at sites on the top of high mountains, several hours hike from the base camp, one needs be prepared for any situation that could arise, like an unexpected storm that requires temporary shelter or a medical emergency. We would set out loaded up with spare tents, food, fuel and stove to cook meals, various snacks, a first aid kit, hammers, chisels and materials for wrapping and documenting specimens. Basically our backpacks were packed full when we set out and even fuller on return. On a few occasions we would find so many good specimens on these treks we would need to make a separate return journey to retrieve all the specimens. If the weather was good, we would sometimes work through the night as it didn't get dark over the summer season, and when one has 24 hours of daylight the actual time of day becomes irrelevant.

At other times, the grim possibility of injury or even death hung in the air, always at unexpected times. I recall at one stage of the journey we were sledging down a steep embankment and the blue ice slope suddenly became steeper than we could handle, and so the rear sledges kept overtaking the skidoo, which was at maximum speed and almost out of control. We were heading to crash straight into a jagged rocky outcrop or even a cliff at the end of the slope, but at the last minute, with me breaking as hard as I could from the rear sledge, the skidoo was able to power away to one side and avoid the imminent crash.

But the most dangerous day in the field was yet to come. Towards the end of the expedition we had been sledging all day to reach the base of a well-known fossil site, discovered by an earlier expedition some 20 years before, at Portal Mountain. It had been snowing on and off, even during brilliant sunny days, the week before as we approached the majestic peaks of the Lashly Range. The thick snow had built up making it easy for our sledges to reach almost record speeds as we traversed some 40 miles that day before setting up our base camp in expectation of New Years Eve celebrations, the arrival of 1992. Our Christmas had been dismal that year as we had separated from our main depot of supplies for a quick sojourn to a site that had turned into a longer stay than expected due to sudden bad weather. We had almost ran out of food, and were down to meagre rations before we could move again and finally relocate our depot. So after we rehitched the two depot sledges full of food to our toboggans and traversed across to Portal Mountain, we were all in high spirits. That night we feasted on some of the extra supplies we should have eaten over the past 10 days but couldn't, and had a few celebratory drinks. We all slept deeply that night as the snow began softly falling once more on our tents.

The next day I woke full of eagerness to get out to the fossil sites on Portal Mountain. The others were tired from the past weeks ordeal and wanted to rest that day. After careful discussion with the team I was given permission to head out alone for a quick exploration of the mountain's lower slopes not far from our base, roughly half a mile away. I loaded my pack with food, supplies and emergency gear as was the routine, then carefully set out across the snow field towards the mountain. As I walked the snow became deeper and was soon up to my waist but I kept ploughing slowly through the snow, edging closer to the rocky slopes full of the promise of discovery.

Suddenly I felt the ground collapse below my feet. In a split second I will always remember vividly, I recall being stopped in mid fall, with my legs dangling below. I thrashed sideways and ended up with my body lying deep in the snow. I slowly got up and looked at the gaping bottomless hole - I had just walked over a crevasse and by sheer luck had managed to move sideways out of the fall. I was shaking, but slowly composed myself and began slowly walking on as the safety of the mountain was not far away, about 100 feet or so. I eventually made it onto the rocks and then had a long rest, quietly shedding a tear or two as the shock of it all rushed upon me.

I had only been away from camp about an hour, but decided that while I was on the mountain I should continue up the slope a bit more to try and reach the fossil bearing layers. The slopes were thick with heavy snow that had fallen over the past 10 days. I carefully chose each footstep to make sure I was on solid ground, walking across the bedding planes. At one stage I came to a steep crossing and had to push my ice pick ahead of me to determine were the solid rock was for my next step. My concentration was down to the ground for this dangerous crossing as I wanted to get to a broader ledge of safe rock. Then I noticed something odd, as small fast bits of ice flew past me down the steep slope. I froze on the spot, and quickly looked up. Bearing down on me was a wall of snow that must have been loosened by my noises or movements. It hit me hard and pushed me over and down the slope and eventually buried me in the small avalanche, with just my head poking up out of the snow. Slowly, carefully, I dug myself out, gathered my wits and walked gingerly to the safety of a nearby wide expanse of bedrock. I lost myself for while in the swamp of emotion that bore down on me, and it was a good hour or so before I moved from that safe spot and contemplated the trip back to basecamp.

I made it back safely by tracing my footsteps carefully, and giving the crevasse a wide berth. On relaying the details of my trip my teammates consoled me, and Brian, our survival leader, gave me a strong neat whisky. I slept very deeply that night.

The next day Brian determined a safe path to get us onto the mountain slopes, by using a 6 foot crevasse pole to test every bit of ground between us and the mountain. In doing so we learned that my tracks the day before had serendipitously passed over 6 crevasses, any one of which could have caused my death, and through nothing else but sheer bloody luck, I survived. I still have a small, fossil fragment of fish bone I collected that day up on the mountain on my way back to camp as a memento to the fragility of life today, and how life endures through time, despite the odds.

That expedition sure had some perilous moments for me personally, but in the end I think the risks were far outweighed by the significance of the discoveries we made. That is what science is all about, that is what drives us to go back into the field again and again, against all odds, to search for new discoveries. That next site could yield something really big that will fill a big piece of the missing puzzle about the evolution of life. And that's what makes it all worthwhile.

John Long's memoir of his Antarctic expeditions -"Mountains of Madness- A Scientist's Odyssey in Antarctica" - is available through Joseph Henry Press, DC)

John's recent book on the prehistory of Antarctica, "Frozen in Time- A prehistory of Antarctica" is available through CSIRO Publishing, Australia.

The Immortal Jellyfish Turritopsis and the Future Evolution of Humans

Shin Kubota

Shin Kubota is an associate professor at Seto Marine Biological Laboratory, Field Science Education and Research Center at Kyoto University, Japan.

Study of the ontogeny of jellyfish from gametes to adults is my daily work that I carry out in the laboratory and field. Such studies may reveal ongoing speciation and the origin of higher taxa. Natural history studies of the diverse animals in Japan, particularly inhabiting the vicinity of the Seto Marine Biological Laboratory, Kyoto University around Shirahama town, Wakayama Prefectue, involves my basic work. I examine organisms' life history, development, distribution, behaviour, ecology, and biogeography. My Cnidarian biological research is focused on (1) systematics based on life cycle and Green Fluorescent Protein (GFP) pattern; (2) systematics and life cycle of a Cambrian cnidarian fossil; (3) marine biological study of immortal *Turritopsis* and ephemeral *Eugymnanthea* and its relatives.

Since my earliest childhood I have been attracted to marvelous animal life, especially marine life due to its enormous diversity. I was born in a seaside town and would play with sea animals through snorkeling and fishing, etc.

I was impressed by Jules Gabriel Verne's *20,000 Leagues Under the Sea* at an elementary school age. In this science fiction future, inventions are anticipated and Captain Nemo's deep thoughts are meaningful in every epoch even for a child. But this novel is very adventurously written and my favorite diverse animals from jellyfish to whales appear here and there.

Accordingly my major work is still now on the adventurous natural history of marine animals. Thanks to the very rich fauna of Japan, most of my (and corroborators') recent biological records (about 450 in number) extending from Cnidaria to Vertebrata (including Ctenophora,

Gastrotricha, Echiura, Annelida, Mollusca, Crustacea, Uniramia, Chelicelata, Echinodermata, and Urochordata, etc.) can be readable to Kyoto University's open access Research Information Repository KURENAI (http://tinyurl.com/kurenairepos).

Jellyfish research, actually, is fascinating, meaningful work. Jellyfish studies have been carried out in our lab since 1911, since our area is a so-called a jellyfish paradise sea. Japan's tremendous number of jellyfish have been described and recorded in many efforts by staff of my lab, allowing me to produce a book in 2014, after introducing each jellyfish species in a weekly article of a newspaper for more than 3 years.

Jellyfish are particularly suitable for biological study since they are clonal animals and in nature many copies exist. We can use a part of copies as our study materials, with no effects to nature. Jellyfish are so graceful in their transparent, tetra-radially symmetrical body plan, spending a complicated life in a water environment. Thousands of species have evolved from ancient times, dating back to the Cambrian, c 5.5 billion years ago (*Early Cambrian Pentamerous Cubozoan Embryos from South China* - journals.plos.org/plosone/article?id=10.1371/journal.pone.0070741).

Extant cubozoans are voracious predators characterized by their square shape, four evenly spaced outstretched tentacles and well-developed eyes. A few cubozoan fossils were known from the Middle Cambrian Marjum Formation of Utah and the well-known Carboniferous Mazon Creek Formation of Illinois. Undisputed cubozoan fossils were previously unknown from the early Cambrian; by that time probably all representatives of the living marine phyla, especially those of basal animals, should have evolved.

Fortunately, microscopic fossils were recovered from a phosphatic limestone in the Lower Cambrian Kuanchuanpu Formation of South China using traditional acetic-acid maceration. We analyzed in detail through computed microtomography (Micro-CT) and scanning electron microscopy (SEM) seven of the pre-hatched pentamerous cubozoan embryos, each of which borne five pairs of subumbrellar tentacle buds.

The microscopic fossils were unequivocal pre-hatching embryos based on their spherical fertilization envelope and the enclosed soft-tissue that had preserved key anatomical features arranged in perfect pentaradial symmetry, allowing detailed comparison with modern cnidarians, especially medusozoans. A combination of features, such as the claustrum, gonad-lamella, suspensorium and velarium suspended by the frenula, occur exclusively in the gastrovascular system of extant cubozoans, indicating a cubozoan affinity for these fossils. Additionally, the interior anatomy of these embryonic cubozoan fossils unprecedentedly exhibited the development of many new septum-derived lamellae and well-partitioned gastric pockets unknown in living cubozoans, implying that ancestral cubozoans had already evolved highly specialized structures displaying unexpected complexity at the dawn of the Cambrian. The well-developed endodermic lamellae and gastric pockets developed in the late embryonic stages of these cubozoan fossils were comparable with extant pelagic juvenile cubomedusae rather than sessile cubopolyps, which indicated a direct development in these fossil taxa, lacking characteristic stages of a typical cnidarian metagenesis such as planktonic planula and sessile polyps.

I have studied jellyfish since 1976, since I noticed cnidarians stand on an oldest node, a root, of the evolutionary animal life tree and, therefore, thought they may reveal a hopeful secret of life hidden in its body. They're simple organisms. We think they are a holdover from when life first began.

Among these smaller jellyfish is *Turritopsis dohrnii.* Out of all the animals in the world, only they are able to reverse the aging process instead of dying. As they age, or when they experience trauma or illness, they can become a polyp again.

Indeed this most miraculous feature is found in only jellyfish among all animal groups: it is rejuvenation. A mature jellyfish can become young, transforming back to the polyp stage, and then again grow into an adult. This would be like a butterfly metamorphosing back into a caterpillar. And, what's more, rejuvenation can be repeated, possibly forever! Such a rejuvenation of jellyfish has been demonstrated by me, at least 10 times, in *Turritopsis* (Kubota 2011; *Miracle Die-Hard Jellyfish Scarlet Medusa & Ephemeral Jellyfish.*

Basic findings for this miracle *Turritopsis* from my 40 years of study are, first of all, there are two morphs in Japanese waters, large northern form and small southern form (Kubota 2005). Thence their mode of reproduction is completely different: the large form broods babies, while small form is a non-brooding morph. At last, they appear as three distinct molecular species (Miglietta, et al. 2007), and some scientific names should be changed, particularly for Japanese species. The life cycle study is most fundamental in biology, I believe.

Furthermore, my other noticeable finding has been when culturing *Turritopsis* an odd colony formed when an old body and a new rejuvenated body connected with each other. The old body is a manubrium, of which the internal part is a stomach for eating and the outer part is gonads for sexual reproduction to produce offspring. Interestingly, the individual's life-span can be extended by this connection, from several months to half a year. This basic body part is destined otherwise destined to die. Initially I thought this important part could rejuvenate. However, this idea is not true for *Turritopsis* and also with many other jellyfish species cultured, including my favorite, *Eutima japonica*, an ancestral type of bivalve-inhabiting hydrozoans.

However, to culture of the immortal jellyfish in the lab is quite difficult. Careful daily care is needed. Too much food supply easily degenerates the polyp. Circulation of water is necessary for healthy conditions, etc. Therefore, daily heartful care is needed. It is easy to demonstrate rejuvenation once for everyone, but very difficult even twice since the polyp is fragile in the laboratory. Patience and steady effort backboned by enthusiasm and passion are required for this success.

It can be conceivable that *Turritopsis* conceals secret of immortality of animal life. Well-known induced pluripotent stem cells-like production processes are possibly taking place within its small body when the jellyfish meets danger of life or aging. Rejuvenation is a human's ultimate dream rather than simple cure of any body parts. Application of rejuvenation mechanism of *Turritopsis* to humans is hopeful. (*see* Will Human Dreams of Immortality Come True Through Jellyfish Research? : Biological and life science studies of immortal and ephemeral jellyfishes (tinyurl.com/owf9q3s) and http://www.kyoto-u.ac.jp/explore/professor/).

Shin Kubota

Video: The Jellyfish That Holds a Key to Immortality- http://tinyurl.com/motherboardjellyfish

The best way to acquire consciousness is to learn any secret of life in nature, and this leads us to evoke respect of every precious life. All existence on the earth is based on the continuation of a long life history of billions of years. This understanding has motivated me to make animal songs, especially for extant colleagues living on the earth including immortal jellyfish (http://www.benikurageman.com). Using simple poems and good melodies, 46 songs I have produced so far, we can read and sing together heartfully, and sometimes I become an Immortal Jellyfishman.

Shin Kubota video lecture on Turritopsis - http://tinyurl.com/ideacityshin

References

Kubota, S. 2005. Distinction of two morphotypes of *Turritopsis nutricula* medusae (Cnidaria, Hydrozoa, Anthomedusae) in Japan, with reference to their different abilities to revert to the hydroid stage and their distinct geographical distributions. *Biogeography*, 7: 41–50.

Kubota, S. 2011. Repeating rejuvenation in *Turritopsis*, and immortal hydrozoan. Biogeography 13: 101-103.
Miglietta, M. P., Piraino, S., Kubota, S. and Schuchert, P., 2007. Species in the genus *Turritopsis* (Cnidaria, Hydrozoa): a molecular evaluation. *J. Zool. Syst. Evol. Res.* **45** (1): 11–19.

http://repository.kulib.kyoto-u.ac.jp/dspace/?locale=en

SK HP: http://www.seto.kais.kyoto-u.ac.jp/shinkubo/index.html
SMBL HP: http://www.seto.kais.kyoto-u.ac.jp/
FSERC HP: http://fserc.kais.kyoto-u.ac.jp/
Books, Songs, TV, Radio, etc.: http://www.benikurageman.com
Yahoo Blog: http://blogs.yahoo.co.jp/skyline2011marine/1491108.html

New Electromaterials for Medical Bionics: The Journey So Far

Gordon Wallace

Gordon Wallace is currently the Executive Research Director at the ARC Centre of Excellence for Electromaterials Science *(www.electromaterials.edu.au/) and Director of the* Intelligent Polymer Research Institute. *He is Director of the ANFF Materials node. He previously held an ARC Federation Fellowship and currently holds an ARC Laureate Fellowship. Professor Wallace's research interests include the discoveries and development of new materials for bio-communications from the molecular to skeletal domains in order to improve human performance via medical Bionics (www.youtube.com/ ACESelectromaterials). He has pioneered the use of 3D bio printing to achieve this (www.youtube.com/ACESelectromaterials). With more than 800 refereed publications, Professor Wallace has attracted some 26,000 citations. He recently published an ebook* 3D Bioprinting: Printing Parts for Bodies *(www.bioprintingebook.com). He has worked with Magipics to create animations describing various aspects of his work (tinyurl.com/ acesanimation). He has supervised 94 PhD students to completion at the Intelligent Polymer Research Institute and currently co-supervisors 30 PhD students. Professor Wallace is an elected Fellow at the Australian Academy of Science, the Australian Academy of Technological Sciences and Engineering, the Institute of Physics (UK) and the Royal Australian Chemical Institute.*

Not all friendships result in collaborations but all collaborations require friendship: trust, integrity, respect. The title of this book is apt. This is not a story about personal discoveries but about the realisation that effective research teams bring about much, much, more than the sum of their parts.

Gordon Wallace

Intelligent polymers[1] and a subset of this, Organic Bionics[2], are topics that have fascinated me for three decades. Whilst I may have coined the terms and refined the concepts with input from others, they originated in a cauldron of collaboration.

Our pursuits have seen the development of new biosensors, controlled drug delivery systems, artificial muscles and the discovery of flexible materials that can store energy in wearable systems. We have adapted these materials to bridge the divide between living systems and technology, enabling effective communication with living cells. On the horizon are implants for nerve and muscle regeneration. We are also targeting implants for epilepsy detection and control.

This journey has taken us into the area of additive fabrication - an approach to manufacturing that allows integration of unconventional materials, and even living components such as cells, into practically useful structures and devices.

I assemble these musings in the midst of a collaborative research tour that will take me to laboratories in Ireland, Italy, the Netherlands and Germany over four weeks. I will also meet with collaborators from the United Kingdom, France, and Finland. Such is the nature of the work that started with my undergraduate training in Physics and Chemistry at Deakin University in Geelong, Australia. Fortunately for me, Prof Alan Bond arrived at Deakin as I was completing my undergraduate degree. A summer research scholarship ignited my interest in electrochemistry, materials and biological systems. Results from this summer work were published in the Lancet (http://www.ncbi.nlm.nih.gov/pubmed/6103422).[3] So the first discovery I was involved in included chemists and biologists tackling a clinical question: can copper Intrauterine Devices (IUDs) generate a suspected carcinogen – malonaldehyde? We developed an analytical method to detect malonaldehyde at low levels and found that the presence of copper in cervical mucous could induce formation of malonaldehyde. This was applied to clinical samples and the results published in the Lancet. The recommendation was that careful monitoring was required.

I found I was pretty well prepared for a research laboratory environment. The practical nature of the Deakin undergraduate degree included lots of lab work with hands on glassblowing (ouch!) as well mechanical and electronic workshop projects thrown into the mix. I had also been fortunate to have lab based summer jobs at the Shell Refinery in Geelong and Queensland Nickel in Townsville. In both I discovered some new practical lab skills from some old hands. I completed my PhD degree with Alan and learnt that my instinctively creative approach needed to be tempered by the attention to detail needed for real science. During these studies we discovered a new approach to on-line monitoring of metal ions. The approach was developed in association with a number of industry partners. The constant stream of research visitors to Alan's labs taught me the importance of working with others. While science, like art, is often a personal journey, so much more can be achieved by working in concert with others.

I was fortunate to obtain a Lecturing position at University College Cork in Ireland almost immediately after submitting my PhD thesis. This took me back to my country of birth. A country my family were pleased to leave in 1972 at the height of "the troubles." We immigrated to Australia by boat. Hence my current aversion to the romantic cruise often proposed by my wife, Vicky. Perhaps it was "the troubles" that catalysed my drive to succeed at something, to do well enough that I could leave that place and maybe take our family to safer ground. The fear of failure and the fact that would mean limited choice in determining my future was always at the back of my mind.

My preference was to succeed at football.

George Best did it from the streets of Belfast – why not I?

My second preference was science.

I had scraped past in the 11+ exams at the end of primary school and won a place at a prestigious grammar school. I knew I scraped in because they started me in form 1D. The bright kids were in 1A. I did crawl through the ranks (there was a "promotion/relegation" system) to be in 3A by the time we left Ireland for Geelong in Australia.

The only good thing I remember about the long boat trip was my dad buying me a guitar at our first stop in Spain. I spent the next few weeks learning to play it. I still enjoy playing and creating music.

Gordon Wallace

Upon arriving in Australia I was sent to Oberon High School where in the final years of secondary school I encountered three teachers who inspired me to do the very best I could in Chemistry, Physics and Maths. I thought then that maybe I should do a medical degree so good marks were essential.

The Cork gig gave me an amazing opportunity to pursue my research interests in new electromaterials, and I already knew that it was biological and medical applications that fascinated me. The famous frog's legs experiments of Luigi Galvani (http://tinyurl.com/animalelectricity) intrigued me. I started to work on organic conducting polymers. While our initial discoveries were around metal ion detection, a natural follow on from my PhD studies, the thought of creating new versions of these materials for biological applications was always at the back of my mind. My Cork masters employed me to run a postgraduate course in Analytical Chemistry. Ironic, given one of my lowest grades in my undergraduate degree was in this subject, a result that almost prevented me entering Deakin's PhD program. I think that near miss gave me the kick up the arse I needed. It reminded me where I came from and that to achieve something big I better bloody work for it.

I did that during my PhD and continued in Cork. Building research links across Ireland – some of which are still highly active today. These include strong links with Prof Dermot Diamond's group at DCU. Current projects include the SwEatch, a wearable watch for analysis of sweat composition for training and medical diagnostics and photoswitchable materials, the properties of which can be activated with light, for diagnostics and for controlling cellular interactions. I also developed a close collaboration with Prof. Malcolm Smyth at DCU and he was to be the one that enabled our first bio molecular experiments with conducting polymers during a sabbatical in Wollongong, Australia. We discovered that antibody molecules incorporated into conducting polymers could effectively interact with their corresponding antigen and generate an electronic signal.

Antibody-antigen interactions have been the basis of a number of electrical signal based biosensing approaches. The molecule to be detected couples to its molecular partner with exquisite selectivity. Approaches used prior to ours involved the development of elaborate chemical procedures to bind the hunter molecule to the sensor surface. Here we thought the trapping of the hunter in the unique mobile electronic environment offered by organic conducting polymers may offer some advantages. This discovery was the first step along the journey to develop organic bionics. The demonstration of the ability to interact with delicate biomolecules and to use this new electronic communication platform to monitor biomolecular events was exciting.

After two years in Cork I was throwing my hat in the ring for positions back in Oz (Australia) just to register interest. One sunny afternoon in my elaborate UCC office I received a call that was to change my life. It was an interview of sorts. Prof Leon Kane Maguire called to chat about an opportunity. I do not think I answered a question other than with a "Yes" or "No." Quite a chatter was Leon.

Anyhow I got the job and have been in the Gong ever since. Leon and I went on to forge an important collaboration. We discovered a simple route to inducing chirality in organic conducting polymers. With organic conducting polymers a molecular dopant is used to stabilize the charged conducting state. By choosing that dopant to be chiral, and with appropriate functional groups to interact with the polymer backbone, chirality could be induced in the overall structure. This is important as many biologically active molecules are chiral (simple mirror images of each other) and determining the presence of or interacting with one form or the other can be used to monitor and influence biological processes.

Leon was head of the Chemistry department, and his support and unwavering enthusiasm enabled me to rapidly build national and international networks. He allowed me the flexibility in teaching responsibilities to undertake necessary travel to build collaborations. We engaged in early trips to China together to recruit PhD students who made enormous contributions to our early advances. During one of those early trips, this time to a conference in Singapore, I met Prof Alan MacDiarmid and Dr Ray Baughman.

Alan was awarded the Nobel prize in Chemistry in the year 2000 for the discovery of Organic Conducting Polymers. Alan served as the Chair of our International Advisory Board for 2 years (2005-2006) and as personal mentor for ten years before that. Alan's enthusiasm for scientific research was unwavering. His support of younger researchers including myself was irrepressible. He passed away in 2007.

Ray remains an active and enthusiastic collaborator. Our most recent collaborative venture involving scientists from 6 countries (USA, Australia, Canada, South Korea, Turkey & China) involved the use of coiled fishing line to produce extraordinary artificial muscles. Our earlier collaborations with Ray included work in the area of organic conducting polymer actuators.

The Wollongong appointment allowed me great freedom and, with Leon's support, great scope to further develop my interests in highly functional electromaterials. So functional that soon the idea of Intelligent Polymer Systems was to emerge and the Intelligent Polymer Research Laboratory (IPRL - http://ipri.uow.edu.au/index.html) was established in 1990. The focus on biointeractions enabled us to discover that organic conducting polymers could be used not just to monitor, but also to control, antibody-antigen interactions Using appropriate electrical stimuli we discovered that antibody-antigen interactions could be made to be reversible. This was used in developing new biosensing technologies and more importantly laying the foundation for our subsequent cellular interaction studies.

Video: Nanostructures for Electromaterial - tinyurl.com/nanostructuresvideo

Facilitated by the DVC Research at UOW (Prof Ian Chubb, now Australia's Chief Scientist), Dr Anthony Hodgson joined our group and in the early 90's his skills enabled us to delve more into biological studies. We demonstrated cytocompatibility with organic conducting polymers and also discovered protocols that enabled living cells to be incorporated into these materials during synthesis. The cells were red blood cells. We facilitated their incorporation through the use of large charged molecules (a polyelectrolyte) that would act as the molecular dopant in organic conducting polymers but also chaperoned the blood cell to ensure it could be incorporated intact.

We published these, what I considered to be, exciting findings. To my dismay they were not received with the fanfare I expected and we stumbled along for a few years. An encounter with Prof Graeme Clark some 10 years later catalysed a resurrection of this whole area of our activity and the concept of *Organic Bionics* was born.

We have subsequently discovered and developed new electrode materials for the Cochlear ear implant (The Bionic Ear) as well as materials that will form the basis of nerve or muscle repair conduits. Graeme introduced me to other clinicians with whom we enjoy ongoing collaborations: Prof Stephen O'Leary (Cochlear Implant projects), Prof Peter Choong (conduits for nerve muscle and bone regeneration), and Prof Mark Cook (implants for epilepsy detection and control). In these projects and others we continue to discover new materials that expand the inventory for bionic devices. Graphene is a notable recent addition. In a recent project with Prof Ric Kaner at UCLA we discovered the world's thinnest carbon electrode (2 layers of graphene) capable of electrically stimulating nerve growth.

We have found the need to couple advances in fabrication with advances in materials science in order to make more rapid progress. So we have expanded our fabrication capabilities within our laboratories developing 3D printing machinery that allow us to print structural biopolymers, bioactive molecules, soft conductors (such as soft materials coated with graphene) and even living cells. In collaboration work with Prof. Peter Choong we have found these 3D printed structures to be of great use in facilitating cartilage regeneration from adipose stem cells. The discovery of bio inks that allow us to print intact living cells has opened up some fascinating clinical opportunities in areas such as cartilage regeneration and has also enabled us to probe fundamental biological processes in 3D printed pseudo tissue. Our initial foray into printing structures containing living cells involved ink-jet printing. Advances in fabrication of novel materials has also enabled the discovery of new structural arrangements to create predetermined drug release profiles. This discovery is part of an ambitious project to provide epileptic drugs directly into the brain, in response to brain signals indicative of an impending seizure, a collaborative project with Prof. Mark Cook (neurologist) at St Vincent's Hospital, Melbourne.

The materials discovery journey continually highlights the need for innovative approaches to fabrication. Numerous attempts to engage the more traditional manufacturing sector revealed that was not the way to go. We started to engage in the development of printing technologies. Initially modifying "off the shelf" ink-jet printers, and then building customised extrusion printers (thanks to the support of Prof Paul Calvert, U Mass), we took up this challenge some ten years ago.

We have gone on to create new printing technologies that cover the nano- micro- macroscopic. We have developed inks to deliver biopolymers as structural materials, proteins and living cells distributed with exquisite precision in 3D. 3D bioprinting has revolutionised our approach to both fundamental and implementation studies.

Video: Printing Living Cells - http://tinyurl.com/printinglivingcells

In parallel my interest in sport continues. I had the privilege of coaching a special group of kids from u11-years of age and culminating in a grand final win in 1st Division at u18's. Well since I could not be the next George Best maybe I could help create one.

I also have the great fortune to work with a highly talented group of Biomechanic Researchers led by Prof Julie Steele. Our interest in wearable electronic textile sensors and actuators (artificial muscles) led us into the development of a knee sleeve. A device with integrated electronic textile human movement sensors was developed, and we discovered that this greatly facilitated landing training for AFL footballers. I got to work with one of the other loves of my life – The Geelong Football Club. The heart of the system is a wearable textile based movement sensor. An electronic textile that works as a strain gauge with a huge linear range – able to measure strain greater than 100% strain (typical strain gauge measure 1-10%).

A similar textile based sensing technology was an essential component for development of the Bionic Bra project (http://tinyurl.com/bionicbraproject). Here the sensors detect breast movement and artificial muscles are used to control it should it become excessive. The UOW media story (http://media.uow.edu.au/releases/UOW184372.html).

Practical implementation of organic bionic devices for implantables or wearables will require innovative energy sources. We have built a strong program around the development of flexible electromaterials - electrodes and electrolytes for wearable energy storage systems. We have discovered that hybrid materials containing graphene provide both the mechanical and charge storage properties demanded for such applications. We have also discovered new electrode structures containing graphene for use in what could become implantable glucose fueled biofuel cells.

Video: The Graphene Revolution - Prof. Gordon Wallace - HotHouse STUFF 18/06/14 - tinyurl.com/graphenevideo

A series of fortuitous events enabled the recent development of a device and fabrication facility on our Innovation Campus. This has attracted some amazing talent from around the globe. Leon passed away in 2011 and never got to see the amazing fabrication facility and the brilliant people it would attract.

My family

No journey like this is possible without the support of others.

My parents instilled in me a great sense of responsibility to our family and to others. As I mentioned above that drove me to succeed at something. Not that they demanded so. Quite the opposite – they wanted nothing more than for all of us kids to be healthy and happy.

My mother was 18 years old when I was born. She worked all sorts of jobs to help provide for the family including a number of years at a cigarette factory. I remember one particular job at a local fruit shop because mum also got me a weekend job there packing bags of potatoes. I noticed they just threw out wooden fruit boxes. I asked the owner could I have them. I chopped and bundled and sold bundles of sticks used to light open fires to people in the neighbourhood. That venture soon overtook the potato packing job in terms of revenue raised, but I retained both – mum did not want to go back to having to pack the potatoes!

My Dad was a fitter and turner who worked various jobs in Belfast and in Geelong when we arrived in Australia. He was proud in a quiet sort of way of any achievements any of us kids brought home.

I have been blessed with a special family of my own. My wife Vicky is the mainstay of that enterprise. With my son Jordan and daughter Eileen they provide the stable environment one needs in this topsy-turvey world of research. They raise a glass to success, they provide comfort in disappointment. They provide a reminder of why we do it.

Why, when whatever the circumstances we remember.

No Defeat. No Surrender.

I am proud of what we have achieved together.

I am pretty sure the excitement and sometimes bitter disappointment of the research journey has helped shape our family values. I can see that the sense of responsibility instilled in us by our parents has been transferred to our children. I can see an awareness to make the most of every opportunity to excel with the skills given to them and return to others who have believed in, supported, and nurtured their dreams.

In closing

Twenty five years ago I would not have thought my pursuit of Intelligent Polymers and Organic Bionics would see me delve into the nano world – discover amazing properties in the nano world and now build new chemistries and machineries that enable us to create structures and devices that retain those properties.

The value of friendship and collaboration building in scientific research is undervalued and under resourced. Research organisations need to structure things to better resource this activity and to acknowledge the importance of it. Despite that lack of organisational support, critical collaborations can grow with time and extraordinary personnel commitment.

Only on reflection do we see the journey would have been very different and much less fruitful without these collaborations. Too often that reflection has been catalysed by a sudden and unexpected loss.

Thank you, Dad. Thank you, Alan. Thank you, Leon.

I will sign off now.

I am 57 years old today, and I consider I have limited years to complete this journey and return to others knowledge and practical outcomes. I owe so much to so many. Thank you for helping me acquire the skills and confidence needed to help build an incredible team of individuals. My role now is more of a playing coach. I love coaching but nothing beats the feeling of pulling the boots on your own feet and getting into the game.

As we travel the research road **We Discover** new things about science and engineering.

We Discover new things about people and the skills required for success.

We Discover new things about ourselves.

Video: Next generation medical bionics: Professor Gordon Wallace at TEDxUWollongong - tinyurl.com/gordontedx

References

1. "Conductive Electroactive Polymers: Intelligent Polymer Systems" Third Edition, Wallace, G.G., Spinks, G.M., Kane-Maguire, L.A.P., Teasdale, P.R. *CRC Press*, Taylor & Francis Group, Boca Raton, 2009.
2. "Organic Bionics" Wallace, G.G., Moulton, S.E., Higgins, M.J., Kapsa, R.M.I. *Wiley-VCH Verlag & Co. KGaA*, Boschstr. 12, 69469 Weinheim, Germany 2012.
3. Malonaldehyde in Cervica Mucus associated with Copper IUD Bond, A.M., Briggs, M.H., Deprez, P.P., Jones, R.D., Wallace, G.G. *The Lancet* 1980, 1, 1087.

Explorations of Heroism: My Journey Toward Understanding the Genesis of Exemplary Behavior

Scott T. Allison

Scott Allison is Professor of Psychology at the University of Richmond. His research program focuses on human belief systems about heroes, villains, legends, leaders, underdogs, and martyrs. He has served on the editorial boards of Journal of Personality and Social Psychology, Personality and Social Psychology Bulletin, and Group Dynamics: Theory, Research, and Practice. He has published nearly 100 articles and several books on heroes including Heroes, What They Do & Why We Need Them, *published in 2011 by Oxford University Press, and* Heroic Leadership: A Taxonomy of 100 Exceptional Individuals, *published by Routledge in 2013. His work has been featured in media outlets such as National Public Radio, USA Today, the New York Times, the Los Angeles Times, Slate Magazine, MSNBC, CBS, and the Christian Science Monitor. He is the recipient of the University of Richmond's Distinguished Educator Award and the Virginia Council of Higher Education's Outstanding Faculty Award.*

The famed comparative mythologist Joseph Campbell once said, "We must be willing to get rid of the life we've planned, so as to have the life that is waiting for us." This bit of wisdom most definitely characterizes my life as a college professor, especially the circuitous path that brought me to the study of heroism.

Studying heroes was not on my to-do list as a young assistant professor. Years ago I was interested not in great people, but in the types of situations that give rise to cooperative behavior in groups. I published a number of experiments that examined the conditions under which people placed their group's welfare ahead of their own individual welfare (e.g., Allison & Messick, 1985, 1990; Samuelson & Allison, 1994). In my first published study, we found that people in large groups consume far more than their equal share of resources than do people in small groups. My colleagues and I concluded that overconsumption decisions may be due as much to ignorance as to selfish motives, because in large groups and societies we often don't know what amount is our fair share. But overall, these early studies did reveal that people tend to be self-serving in the way they distribute resources to themselves and others in groups.

Early in my career, I was struck by the ways in which subtle variations in the environment could lead people down the path of either selfishness or selflessness (Allison, McQueen, & Schaerfl, 1992). In one series of experiments, my students and I placed human participants in groups and asked them to distribute resources to themselves and to other group members. There were two conditions in the study. In one condition, the shared resource consisted of small building blocks, which were very easy to count and to divide equally. Each block was worth $2. We called this condition the *partitioned* resource condition, as the resource units (the blocks) were discrete, countable units. In another condition, the resource consisted of sand, which participants needed a small shovel to scoop out and thus was difficult to divide equally. Each pound of sand was worth $2. We called this the *non-partitioned* resource condition.

We hypothesized that because it was easier to divide the resource equally among group members in the partitioned resource condition than in the non-partitioned condition, we would get more violations of the equality rule in the non-partitioned condition. And that's exactly what we found. Perhaps more importantly, the violations of equality in the non-partitioned condition were in a self-serving direction, with participants consuming more than their equal share in the non-partitioned condition. This pattern was especially prevalent in large groups.

This program of research wasn't quite heroism research but it was a close cousin, focusing on the factors that tend to make people behave badly – or well – in group settings.

In 1991, I found myself teaching a "great books" humanities course to first-year students at the University of Richmond. The course was multi-disciplinary and multi-cultural in its emphasis, and it required students to read such books as Shakespeare's *Romeo and Juliet*, Plato's

Symposium, Darwin's *Origin of Species*, the *Analycts* of Confucius, Naguib Mahfouz's *Fountain and Tomb*, Orhan Pamuk's *The White Castle*, and many other great texts from around the globe. What most caught my attention were the two epic stories on the course syllabus: *The Epic of Sundiata* told by the Malinke people of Africa, and the epic novel *Monkey* (also known as *Journey to the West*) written by Wu Cheng'en during China's Ming dynasty.

These two epic adventures were composed at different points of time in human history, and in different parts of the world, and yet they bore a striking resemblance to the two great western epic stories I had read in high school and in college, namely, the *Iliad* and the *Odyssey*. The *Epic of Sundiata* tells the story of the hero Sundiata Keita, the founder of the Mali Empire. Born an ugly hunchback, Sundiata was prophesized to become a great ruler of the Mali people. The existing king felt threatened by this prophecy and thus banished Sundiata from the kingdom, but years later Sundiata returned to defeat the king and establish the great empire. In *Monkey*, a brave young pilgrim named Tripitaka must travel to strange faraway places to retrieve sacred information needed to enlighten the entire Chinese people. Tremendous courage, wisdom, and virtue are needed by Tripitaka to accomplish this objective.

People's fascination with old dead legendary figures caught my attention. Nearly every psychological theory I had encountered was centered on people's fascination with *living* people, not dead people, and so I sensed an opportunity to study how human beings perceive and evaluate the dead. This led my colleagues and I to write articles on the *death positivity bias* (http://themonkeycage.org/2009/03/want_to_be_a_better_leader_die/) – the tendency of people to evaluate the dead more favorably than the living (Allison, Eylon, Beggan, & Bachelder, 2009). When we conduct these studies, we give human participants information about a person – her name, the town she lives in, her job, her hobbies, and her family. In one condition, the person is described as recently deceased. In another condition, she is described as still living. Our findings consistently show that participants give more elevated evaluations of her in the dead condition than in the living condition. If she is believed to be dead, she is rated as more moral, more inspiring, and more heroic.

My interest in evaluations of the dead also led to our discovery of the *frozen in time effect* (http://psp.sagepub.com/content/31/12/1708.abstract) – people's tendency to resist changing their evaluations of the dead even when new information surfaces that challenges that evaluation (Eylon & Allison, 2005). To study this effect, we had participants read about a person who was described as either alive or dead, and then we told participants that new information had surfaced about this person. The new information was either positive (e.g., he gave large amounts of money to charity) or negative (e.g., he illegally dumped toxic waste into streams). Our results showed that our participants were hesitant to change their initial opinions of the person when they believed the person to be dead. But when they thought the person was alive, this positive or negative information significantly changed their earlier evaluations.

What's important to recognize is that I would never have studied these death-related phenomena had I not decided to take a job at Richmond, which is a small, liberal arts college. The great books course offered at Richmond is typically only taught at smaller colleges, and reading and teaching from these great books inspired me to study dead heroes and how these heroes achieve such revered status.

Plain old good luck then came my way. In 2005, my dear friend and colleague, George Goethals, who had toiled for decades at Siberia-like Williams College in Massachusetts, decided to move south and join me on the faculty at the University of Richmond. Goethals came with an expertise in leadership and an impeccable scholarly record. He and I had collaborated in Santa Barbara back in the mid-1980s while I was a graduate student at the University of California. At that time, Goethals was visiting Santa Barbara while on leave from Williams, and he, David Messick, and I embarked on a collaborative project that, on the surface, would seem to have no connection to heroism at all. We set our sights on understanding self-serving biases in social judgments.

Yet somehow, there was indeed an indirect connection to heroism, although we weren't consciously aware of it at the time. Looking back at our 1980s collaborative work in Santa Barbara, I should have realized that some day Goethals and I would surely write about heroes. The first paper we published together, along with David Messick, was inspired by one of our heroes, the boxer Muhammad Ali. We were always fascinated by Ali's influence and leadership outside the ring, particularly his role in making race relations change in the United States. Ali was always his own man. He insisted on being called Muhammad Ali rather than what he referred to as his slave name, Cassius Clay. At first the media refused to go along. But as we know from his long boxing career, Ali never quit. Eventually sports writers and broadcasters recognized that he was right to insist that they call him what he wanted to be called. He led the way for many, many more African Americans to use names that reflected their pride in their racial identity. There was no doubt that he was the first, and that he led the way.

As George Goethals and I tried to identify the qualities that made Ali an effective leader to a largely hostile white establishment, we focused on his wit and his obvious linguistic intelligence. We remembered that when Ali was once asked whether he had deliberately faked a low score on the US Army mental test, so that he could avoid the draft, he mischievously quipped, "I never said I was the smartest, just the greatest" (McNamara, 2009). That self-characterization led us to research some of the limits on people's self-serving biases. The result was our *Social Cognition* paper, "On being better but not smarter than other people: The Muhammad Ali Effect" (Allison, Messick & Goethals, 1989).

The Muhammad Ali effect reflects Ali's intended meaning behind his explanation for failing the army mental test. The effect refers to people's tendency to believe they are better than others, in a moral sense, but not necessarily smarter than others. Previous research had firmly established that people believe they are better than others, and the assumption was that this belief held true in the domains of both morality and ability. Goethals, Messick, and I challenged this assumption by asking people to rate how they compare to others on both moral and intellectuals dimensions, and we found clear evidence for the Ali effect: People believe they are more moral than others, but they are far less likely to claim intellectual superiority.

At that point neither Goethals nor I had turned to studying heroism or leadership or the connections between them. But we were inching closer in that direction. I joined the faculty at Richmond in 1987 and continued to conduct work focusing on pro-social behavior in groups, examining the conditions under which people place their group's well-being ahead of their own individual interests. Goethals, meanwhile, returned to Williams and was publishing some great work on group goals, social judgment processes, and eventually leadership.

When Goethals was coaxed to join the faculty at Richmond in 2004, he and I renewed our collaboration, this time focusing on the *underdog effect* – the tendency of people to root for disadvantaged entities in competition. This research was borne out of our earlier interest in such diverse heroes such as Muhammad Ali, Sundiata, and Odysseus, all of whom somehow overcame the most terrible adversity to achieve greatness. Goethals and I embarked on a research program exploring people's love for underdogs (Kim et al., 2008).

Our underdog research was great fun to conduct because it surprised us so much. Sure, we found that people love underdogs, but we were shocked to discover how ephemeral this effect can be. For example, when we asked participants whether they would root for Wal-Mart to succeed or for Mom and Pop's small electronics store to succeed, they almost universally preferred to see Mom and Pop prevail. But when asked where they would shop for electronics, they almost universally chose Wal-Mart over poor Mom and Pop. People abandon underdogs when supporting those underdogs may cost them money. We concluded in our article that "the underdog effect is a mile wide but an inch deep."

Gradually our underdog research evolved into work examining triumphant underdogs who became exemplary leaders and heroes. Our interest in underdogs, Goethals' exceptional scholarship on U.S. Presidents, and my own research on people's reverence for the dead (Allison et al., 2009), all eventually led to the books and articles on heroes that Goethals and I have written today (Allison & Goethals, 2008, 2011, 2013, in press; Goethals & Allison, 2012).

Our first book on heroes, *Heroes: What They Do & Why We Need Them* (Allison & Goethals, 2011) addressed the psychology of constructing heroes in our minds as well as the path that great heroes take when they perform their heroic work. Our mental constructions are based on schemas, scripts, and archetypes that provide us with a mental checklist of the attributes we use to identify heroes. We identified the "Great Eight" characteristics of heroes: *strength, intelligence, warmth, charisma, inspiration, reliability, selflessness,* and *resilience*. This book also describes the hero's path as first outlined by Joseph Campbell, the great comparative mythologist. According to Campbell, heroes are called on a journey; they enter a dangerous unfamiliar world; they encounter people who occupy important social roles, such as father figures, lovers, mentors, and sidekicks; they overcome adversity by discovering a missing inner quality; and they return to their original world transformed by their experiences.

Although scholarship on leadership, particularly Howard Gardner's (1997) *Leading Minds*, was always important in the way we thought about heroes, our general exploration of the psychology of heroism diverted us from focusing on the connections between leadership and heroism. Those connections were explored more fully in our review article in *Advances in Experimental Social Psychology* (Goethals & Allison, 2012), where Goethals and I proposed a conceptual framework for understanding heroism in terms of the *influence* that heroes exert. Heroes, we argued, vary in their *depth* of influence, their *breadth* of influence, their *duration* of influence, and the *timing* of their influence.

Central to our analysis of heroic leaders is the taxonomy that Goethals and I developed to understand different types of heroism. We identified 10 categories of heroes: Trending Heroes, who are in the process of either becoming heroes or losing their heroic status; Transitory Heroes, whose heroism is fleeting; Transparent Heroes, who invisibly go about their heroic work behind the scenes; Transitional Heroes, whom we outgrow as we mature; Tragic Heroes, who self-destruct; Transposed Heroes, who quickly move from hero to villain or from villain to hero; Transfigured Heroes, whose heroism we exaggerate; Traditional Heroes, who follow Campbell's classic hero's journey; Transforming Heroes, who forever transform themselves and society; and Transcendent Heroes, whose heroism is so great and extensive it exceeds even the transformative category.

But there was clearly much more to consider. This became increasingly clear in 2010 when we started to blog about heroes (blog.richmond.edu/heroes). Within four years we had written more than 150 hero analyses, attracting over a quarter of a million visitors to the blog. One hundred of our hero profiles were included in our book on *Heroic Leadership* (Allison & Goethals, 2013). Profiling so many great individuals made it increasingly clear that all of our heroes were also leaders. They might not fit traditional leader schemas, or people's implicit theories of leadership, but they were clearly leaders in the sense that Gardner defined it in 1997. Either directly or indirectly, through face-to-face contact or through their accomplishments, products and performances, heroes influence and lead significant numbers of other people.

While I will leave it to others to assess the significance of my body of work on heroes, I do wish to share two observations about the history of my ongoing research on heroism. These reflections speak more to the path I have taken in my work than they do to any destination I have reached. My first observation is that I have benefited from researching the concept of heroism from multiple paradigmatic angles and methodological perspectives. For over thirty years I've looked at selfless behavior using case studies, interviews, surveys, experimentation, dispositional analysis, and contextual approaches. Philosopher William James once wrote that science is best served when scientists not only remain open to fresh perspectives, but actively seek them out. James believed that a single perspective offers but a mere, limited slice of the world (James, 1909/1977). Adopting multiple scientific perspectives expands what one can observe and thus can learn about a phenomenon (James, 1899/1983b). I have found this idea to be certainly true in my study of heroism.

My second observation relates to the Joseph Campbell quote that began this essay. We may think that we can plan how our careers will unfold, but in reality outside forces are always at work that have a far more powerful effect on our professional lives than anything we could ever imagine. What exactly are these "outside forces"? They are the influential people, resources, circumstances, luck, and zeitgeist which are forever lurking and shifting around us. For me, these factors included David Messick's willingness to serve as my advisor in graduate school, George Goethals' decision to choose Santa Barbara as the location for his leave in 1985, my choice to work at a small liberal arts school like Richmond which offered that "great books" course, Richmond's school of leadership offering a position to Goethals in 2004, and many, many more happy chance events.

The serendipitous events that shape our lives are inescapable. During my career, I have been swept and swayed by these influences and have tried not to fight them but to embrace them. These ever-present and ever-changing forces underscore the truism that nothing we can plan in life is ever as special as the unintended route we ultimately take. Dan Gilbert, the eminent social psychologist at Harvard University, was once asked, "What's the key to success?" His immediate reply: "Get lucky. Accidentally find yourself at the right place and the right time." The idea here is that while we'd like to think we are the architects of our own destiny, we are more the product of forces beyond our control than we would like to think. Gilbert later went on to explain this idea more fully in his best-selling book entitled, appropriately enough, *Stumbling on Happiness* (Gilbert, 2007).

"Serendipity," wrote scientist Pek van Andel, "is the art of discovering an unsought finding." Many unsought events had to come together for George Goethals and me to embark on our exploration of heroes. The beautiful orchestration of unpursued circumstances led to the books and articles on heroism that we published (Allison & Goethals, 2008, 2011, 2013, in press; Goethals & Allison, 2012; Goethals, Allison, Kramer, & Messick, in press). The wondrous thing about serendipity is that it has our best interests in mind, as long as we trust it. We need only remain open to receiving, and capitalizing on, the unexpected gifts and opportunities that sly happenstance throws our way.

References

Allison, S. T., Eylon, D., Beggan, J.K., & Bachelder, J. (2009). The demise of leadership: Positivity and negativity in evaluations of dead leaders. *The Leadership Quarterly*, *20*, 115-129.

Allison, S. T., & Goethals, G. R. (2008). Deifying the dead and downtrodden: Sympathetic figures as inspirational leaders. In C.L. Hoyt, G. R. Goethals, & D. R. Forsyth (Eds.), *Leadership at the crossroads: Psychology and leadership*. Westport, CT: Praeger.

Allison, S. T., & Goethals, G. R. (2011). *Heroes: What They Do and Why We Need Them.* New York: Oxford University Press.

Allison, S. T., & Goethals, G. R. (2013). *Heroic Leadership: An Influence Taxonomy of 100 Exceptional Individuals*. New York: Routledge.

Allison, S. T., & Goethals, G. R. (in press). The heroic leadership dynamic: Constructing and executing the most exemplary form of leadership. In Goethals, G. R., Allison, S. T., Kramer, R., & Messick, D. (Eds.), *Core concepts in the psychology of leadership*. New York: Palgrave Macmillan.

Allison, S. T., & Messick, D. M. (1985). Effects of experience on performance
in a replenishable resource trap. *Journal of Personality and Social Psychology, 49,* 943-948.

Allison, S. T., & Messick, D. M. (1990). Social decision heuristics and the use of shared resources. Journal of Behavioral Decision Making, 3, 195-204.

Allison, S. T., McQueen, L. R., & Schaerfl, L. M. (1992). Social decision making processes and the equal partitionment of shared resources. *Journal of Experimental Social Psychology, 28*, 23-42.

Allison, S. T., Messick, D. M., & Goethals, G. R. (1989). On being better but not smarter than others: The Muhammad Ali effect. Social Cognition, 7, 275-296.

Eylon, D., & Allison, S. T. (2005). The frozen in time effect in evaluations of the dead. Personality and Social Psychology Bulletin, 31, 1708-1717.

Gardner, H. (1997). *Leading minds — An anatomy of leadership.* Harper & Collins, London.

Gilbert, D. (2007). *Stumbling on happiness*. New York: Vintage.

Goethals, G. R. & Allison, S. T. (2012). Making heroes: The construction of courage, competence and virtue. *Advances in Experimental Social Psychology, 46*, 183-235.

Goethals, G. R., Allison, S. T., Kramer, R., & Messick, D. (Eds.) (in press). *Core concepts in the psychology of leadership*. New York: Palgrave Macmillan.

Goethals, G. R., Messick, D. M., & Allison, S. T. (1991). The uniqueness bias: Studies of constructive social comparison. In J. Suls & B. Wills (Eds.), *Social Comparison: Contemporary theory and research* (pp. 149-176). New York: Lawrence Erlbaum.

James, W. (1977). A pluralistic universe. In F. H. Burkahradt, F. Bowers, & I. K. Skrupskelis (Eds.), *The works of William James*. Cambridge: Harvard University Press. (Original work published 1909)

James, W. (1983b). What makes a life significant? In F. H. Burkahradt, F. Bowers, & I. K. Skrupskelis (Eds.), *The works of William James: Talks to teachers on psychology and to students on some of life's ideals* (pp. 150–167). Cambridge: Harvard University Press. (Original work published 1899)

Kim, J., Allison, S. T., Eylon, D., Goethals, G., Markus, M., McGuire, H., & Hindle, S. (2008). Rooting for (and then Abandoning) the Underdog. *Journal of Applied Social Psychology, 38,* 2550-2573.

McNamara, M. (2009). *Muhammad Ali's new fight: Literacy.* Retrieved from http://www.cbsnews.com/2100-18563_162-2207050.html on June 15, 2012.

Samuelson, C. D., & Allison, S. T. (1994). Cognitive factors affecting the use of social decision heuristics when sharing resources. Organizational Behavior and Human Decision Processes, 58, 1-27.

Wildlife Cam

Mehdi Bakhtiari

Mehdi is a camera engineer. In 1996, he designed the Crittercam for National Geographic Society.

Intro

My goal was to collaborate with scientists for a better understanding of the lives of animals in the remotest places of our planet, where no human observation has been possible thus far. From Narwhal under the Arctic Ice, lions in Africa, to pigeons in the sky, I have been involved in more than 2000 deployments outfitting over 50 different animal species with cameras.

Since childhood, I was interested in engineering and inventions. I also loved animals. I was fascinated by dismantling objects with a mechanical/electronic look, always eager to get my hands on them just to reassemble and reconfigure them. I was convinced that to be taken seriously in my field, I had to obtain an advanced University degree in Engineering.

As such, I enrolled at John Hopkins University where I was awarded a Master's Degree in Engineering. My head was filled with, what I considered, great ideas and inventions; but I soon realized that the development of these ideas were much harder than I originally had thought. The lack of necessary funds to develop my ideas were a major stumbling block. While frustrated and trying to figure out my next move, I was informed that National Geographic was looking for part time engineers to develop a camera that could film while mounted on the back of a whale.

It was this employment along with cooperation that paved the way for my increased interest for wildlife, hence directing the rest of my life. Despite being overqualified for the position, I agreed to a minimal salary. I was really dedicated to the idea of making such a camera. In just about six months, I managed to create the first-ever camera that could ride on the back of a mighty sperm whale, which actually worked. We recorded footage and got our camera back. It was an AMAZING moment to remember.

National Geographic discovered my talent and the forthcoming of a new horizon. They established a new department called “Remote Imaging.” I found myself managing this department for over ten years, during which I created different types of critter cameras that we deployed onto 50 various species of animals. It was a pioneering achievement that no one had ever done before.

Video: Crittercam Sperm Whales - http://tinyurl.com/crittercamspermwhales

Although we witnessed considerable progress in this field, given the limits in the existing technology, we could not possibly have recordings exceeding a few hours. Sadly, the Crittercam’s far-reaching ability fell short. Yet, this camera tool was not viable enough for the science. I always felt that this camera was not doing the job of which it was capable. We always desired longer video recording of our deployments, hence developing a serious scientific tool. Science was really in need of more feedback enabling the creation of more vital data. Our initial approach was to design and develop a universal standard multipurpose camera, but in reality, I was soon convinced that cameras had to be customized for every individual animal species and its related research scientist.

Going out on my own

In 2007, I decided to establish my own company (ExEye LLC), by customizing cameras according to animal species. I was convinced that this approach would be more appreciated by the scientific community and would lead to better results.

Despite all these, often interesting, eye-opening, imagery that existed in traditional wildlife documentaries, I always believed that none of the so-called documentaries properly portrayed natural history or nature as it truly is. With the very presence of a cinematographer, the possibility of disturbing the animals is very likely which in turn leads to a changed natural behavior. Wildlife filmmakers have been busy getting amazing imagery that is only a snapshot of an animal's life and is often a disturbed one. Documentaries are also limited by length, and filmmakers often aim at higher ratings, rather than the scientific study of the subject.

As such, I invented the first ever long recording camera and named it "The Wildlife Camera," with the goal of outfitting wild animals with cameras capable of hundreds of hours of recordings. I purposely designed it small and light so that scientists could mount the camera on an animal for a period up to a year, to record the true natural behavior of an animal once it gets used to the mounted camera.

This instrument is a data logger that uses a built-in computer to collect different data such as video, audio, 3D acceleration, 3D velocity, 3D compass, ambient temperature, latitude, longitude, pressure, etc. It also correlates them all on an interactive display for ease of interpolation.

Even though the electronic design is very challenging, it is only a small part of the whole design. After designing a camera with a built in computer to control the sensors and with enough memory for hundreds of hours of recordings, it was time to design a power source that was capable of supplying the camera for the required time.

In view of the available technology, we can record up to 240 hrs. of video . For most projects, a 10 day (24hrs X10days=240 hrs.) straight video is not the best solution. We need to control the camera and be selective in order to prolong the deployment from 10 days to months and years. In order to make this a valuable instrument for science, we also need an easy way to monitor and correlate all this data from each deployment.

To overcome this challenge we utilize a microcontroller during the deployment and analyze the program for post deployment, to precisely collect the data under specified conditions. For example, if we are interested in animal diet, we will program the camera accordingly or record only while the animal has its head down or while are running or hunting. By using our analyzed program, we can go through thousands of hours of data, and select the related video without having to watch the whole video.

Utilizing this program has great potential to increase the productivity of the entire instrument research field. My next plan is to share this technology and program with all scientific institutes and universities in order for everyone to share the data online for future generations to study and comment.

We are always in the process of design and development. Every instrument is custom designed with the latest technology for various species in order to fit each animal the best way possible. It should have the least possible effect on animal behavior, while it should be also powerful enough to collect data for thousands of hours. Every animal has its own limitations. For some animals, like dolphins, the camera not only needs to be pressure resistant and hydrodynamic, but also lightweight, small in size with a very soft attachment, so as not to affect the dolphin's normal behavior. For sperm whales, in contrast, the camera can be bigger. It has to be light enough to enable a rapid deployment, but strong enough to withstand over 5000 ft. of water pressure, sensitive enough to see in pitch dark at depth, and buoyant enough (have negative weight underwater) to surface after released from the animal. But a polar bear camera has to be light and low profile and strong enough to withstand polar bears and -40 degree temperatures. An eagle camera needs to be very light, but strong enough to withstand the bird's beak.

The real challenge was the mechanical part (Attachment/Release). Usually following every successful deployment everyone is happy and celebrates their achievement, while for me, the real challenge begins after the deployment. From then on, everything has to be done perfectly in order to recover the camera. If the attachment is not done properly, it will make the data useless, and we may even endanger the life of the animal. And, without release or recovery, the project will fail.

This is why I love this job! There is never a dull moment in my work. I am challenged all the time; every moment is a new beginning for me.

Usually, a project starts with a question from the scientific community. A scientist contacts me and discusses a concern or a question regarding an animal. We usually discuss the issue, and I try to find the best method to attach the camera to the specific species. Once done, I program the camera to trigger at a certain time or situation according to the camera's sensors. For example, I program the camera to trigger only if the animal is at a certain angle, location, and time of the day. This capability allows us to concentrate all our video recordings on a specific issue. Regardless of having solved the first issue, after watching the video, we are always faced with more questions than answers.

We decided once to attach a camera on Narwhals in North Canada to find out why some of them have tusks while others don't, as well as, for what they are used. But shortly after our first deployment, we realized they also swim upside down and swim very close to the bottom. This has caused us to continue our research for years. The gift that keeps on giving!

This invention led to the realization of my dream. By just purely watching the hours of the footage that animals shot of their own lives, we no longer are in need of making up a natural history story. For the first-time-ever, we have been able to see the real story of wildlife; a story that Science could actually analyze and draw conclusions from, which would add to our natural history archive.

Timing is everything with history

Today, what Wildlife Cam is accomplishing for Science is similar to what oral history did to explain ancient civilizations, as well as what written words, films and photography have done for more recent human history.

At the present time, we are at a pivotal technological point in our evolution that will reveal and allow us to discover things we could have never known, prior to the development of Wildlife Cam. GPS, satellite transmitters, VHF tracking systems and other scientific tools might have revealed where animals go, but what behavior they are actually exhibiting when they are in remote places, while not observed by humans thus far have been nothing more than assumptions -- that is until now.

The surprise, Wildlife Cam first deployments had emotional impact

My goal with Wildlife Cam was to allow scientists to have access to an unedited, undisturbed visual of animal life, from the animals' point of view, to further knowledge. The hours animals spend while sleeping and being inactive are just as important to Science as the exciting moments when they are hunting.

But what I did not expect, was just how much these videos affected the viewer on a human level. These recorded videos were very emotional -- thus making them really powerful.

With the first project using my new Wildlife Cam, we attached 35 cameras onto 35 individual caribous. The results made me realize how powerful these cameras were, as compared to anything I had ever done in the past. The first generation of Wildlife Cam designed for caribou focused solely on the caribou's diet. The camera was sampling video during the day, while they were eating the most. Cameras were set to record for different parts of the day, during different seasons of the year, so as to analyze their eating habits at those times.

Unlike the prior generation, these cameras could record for tens of hours. By distributing these long videos into smaller clips, with an onboard computer, I was able to record up to one year of an animal's life.

It is very amazing to watch the video from each deployment to another. One could hardly step away from the video, as they would want to know what will happen next in that animal's life.

During one deployment, we attached the camera on a pregnant female caribou in the winter. We could see how hard it was for this mother to find something to eat when everything was covered by snow and the temperature was well below zero.

Then, at the end of Winter, the female caribou went into labor. If you just look at the GPS tracking the DATA, you can tell the animal is not moving as much as before. You could assume that she was stopping, but you were not sure. Then, you look at the captured video by female caribou, you can watch as she gives birth to a stillborn. At first, you are relieved and happy for her, but then you realize the baby is dead, and you feel sad for her and get depressed. I was amazed that the raw footage could evoke such emotions without any editing.

For the next three days, the caribou's baby doesn't move, and it is gut wrenching to watch the mother attempting to revive it without eating anything herself. After the third day, the mother accepted her baby's death. With the season changing to spring, the snow melted and the water forced her to move on.

At the same time spring, allowed the caribou to find enough vegetation to feed on. I found myself sitting on the edge of my seat, worried that this caribou might die in the winter, and elated once she found food in the spring. It brought tears to my eyes, because I was so relieved that she could finally eat. It was so powerful – as if I was concerned for my own child or a close friend. The emotion I garnered from watching the video truly caught me off guard; and I found myself wanting to watch more.

By following an animal for months on end, and watching every move she makes, you feel like you are part of every reaction and every decision making process. Not only do you learn about the animal, but also get emotionally involved in the plight of a single individual's life. You begin to care tremendously for that animal and have an increased respect for their lives and the species as a whole.

Some discoveries

From our caribou deployment, we were able to discover how a caribou can find food in the Arctic in the middle of winter; they are not starving, as Science once thought. We can even see what kind of vegetation they are interested in. Objectives of the project originally was to understand the diet of caribou, but once the scientists saw the capabilities of the Wildlife cam, they expanded their objectives covering the caribous' diet to habitat selection by type of behavior, such as feeding, resting, traveling, and by each season. The videos not only give us much finer scale data, but they help us correlate the data with visual reference. Even the way we saw the caribou wake up after being tranquilized, allowed scientists to adjust the dosages of the tranquilizer, so as not to affect the caribou more than necessary in the process.

With the Polar Bear deployments, we saw for the first time what bears were doing in Alaska, as the sea ice was melting right beneath their feet. They were chasing seals under the ice and socializing with one another; something Science had no idea they were doing in these remote areas. Scientists used to think that berry eating alone in the ice free seasons could not provide enough nutritional benefit to the bears. But, in the fall, we witnessed bears eating berries non-stop.

Video: Polar Bear - POV Cams (Spring 2014) - http://tinyurl.com/polarbearpov

We can also refer to the upside down swimming behavior of free-ranging narwhals, from the Narwhal deployment. We could document how these unusual animals swim upside down on the bottom of the ocean, and how they utilize their tusk.

Video: Upside-down swimming behaviour of free-ranging narwhals - http://tinyurl.com/upsidedownnarwal

From years of deer deployments, we can see how these animals are protecting themselves and how they get attacked by wolves and mountain lions.

Video: Mule Deer - http://tinyurl.com/muledeerpov

The Sperm Whale POV shows how sociable these animals really are. We can grasp from every deployment, how animals communicate with each other. Almost in all cases, camera will come off prematurely, by the help of a second whale. You feel like they call each other to broom one another!

From Narwhal deployments to Deer cams, outfitting turtles, grizzly bears, and sperm whales, the Wildlife Cam has proven itself to be a uniquely amazing tool that has helped answer questions about wildlife behavior, which until, now scientists could only make assumptions about.

Video: Crack the Blue Whale Code (http://tinyurl.com/bluewhalecode)
Video: Bowhead Whales POV (http://tinyurl.com/bowheadpov)

Game changer

I started this endeavor so as to invent an instrument to gather eye catching images. To be able to create documents that impact both science and human emotion. Today, Wildlife Cam is a game changer for science as well as the public, to acquire new knowledge for science and gain more compassion for animals as a whole.

During these years, while I spent most of my time with wild animals, from the North to South Pole, I learned to have respect for animals. This respect has made me realize that my work is not about us - humans - but is about them and improving their lives. At the beginning, I was trying to catch amazing image shots to appease my boss. As time went by, I learned I was meant to be working for animals, and they were my boss. I respected them very much. I would not do anything unless I see it as beneficial to them. With today's technology, I believe we should be able to get the most data out of each deployment. I don't think that a few hours of deployment are any more acceptable. With the addition of sensors, scientists are really making strides on being able to compare data, blood work, movement, and the actual video of what animals are doing to answer behavioral questions that have baffled scientists for years. To come together with technology, and with the help of animals outfitted with Wildlife Cams, I believe that we have just entered an amazing phase in revealing natural history science and behavior; an innovative phase, where we will outfit every wild animal species on our planet with Wildlife Cams, and where through the animals'point of view, we will learn about their real lives.

Childhood's End: The Imminent Discovery of Alien Life on a New World

Richard K. Barry

Richard K. Barry was the lead power system engineer on the Rossi X-Ray Timing Explorer, the Transition Region, and Coronal Explorer, and the Wide-Field Infrared Explorer spacecraft at NASA Goddard Space Flight Center. He then transferred from power systems engineering to science instrument design and test and has developed instrumentation for the Spitzer Space Telescope, the James Webb Space Telescope, Microwave Anisotropy Probe, and Hubble Space Telescope. Richard conducted instrument modeling and the first science with NASA's pathfinder exoplanet observatory: the Keck Interferometer Nuller. This interferometrically combined the light-gathering power of the two largest optical telescopes in the world on a long baseline to permit very high spatial resolution images of distant objects. He has conducted exoplanet transit research using NASA's Deep Impact probe and Spitzer Space Telescope. He has studied gravitational microlensing using the warping of spacetime to detect exoplanets and co-discovered two new planets from the Canopus Observatory in Tasmania and Deep Impact probe in Earth-trailing orbit. He is currently a project scientist for NASA'S Wide-Field Infrared Explorer, a flagship mission to characterize Dark Energy and to complete the census of all possible exoplanets in the Milky Way Galaxy.

My position as a NASA astrophysicist came in a roundabout way. My experience proves that you don't have to follow a particular path to become a research scientist. What matters is tenacity and self-confidence in the face of adversity.

An abysmal student in high school, I failed out of math and almost every other subject in my sophomore year. I avoided challenging classes and graduated somewhere near the absolute bottom of my class. I had no plan for college. I had no plan of any kind. My home life was a disaster; with divorced parents and a stepmother who openly despised me. I slept in the basement of my father's home as I worked at the only job I could find, as a laborer at the Diamond Crystal Salt Factory in St. Clair, Michigan. All day I threw 100-pound bags of salt from a wooden slat conveyor to a pallet. I would fill one pallet, and then a forklift would quickly replace it with another. Lifting salt bags at a rate of about one every 10 seconds for 8 hours, I would throw about 2,300 bags- 115 tons of salt- on a typical shift.

I recall staggering out of the plant at twilight, covered from head to foot with sweat and salt dust and looking up at the stars. I loved them since I was a boy growing up on a farm, but I never thought I could study them professionally. That seemed too distant a dream - an impossible, unscalable mountain for someone like me. A robot eventually took my job, and having no other options and no money, I enlisted in the U.S. Air Force as an F-16 technician. My 5-year tour took me to many places, including Okinawa, Japan, where I fell in love with scuba diving and dove most of the major reefs in the Ryukyuan archipelago in the East China Sea. On one summer night dive, I turned off my lights and watched my bubbles rise through banks of tiny, startled phosphorescent sea creatures toward a watery full Moon. This otherworldly vision led me to recall my old dream, and the next day I enrolled in night school, taking every math class I could, one at a time - starting over with high-school algebra.

I drilled deeply into each class - seeking mastery, building self-confidence. Amazingly, I was able to excel in every math subject all the way through advanced calculus and partial differential equations. Eventually, the college extension program on Okinawa ran out of available math classes and I had to petition to attend the University of Ryukyu with Japanese students. In this setting I got my first 'B' since high school. I remember how it stung! Looking back, I realize now that the only thing that was different between the utter failure of a high school student and what I had become was having a real vision for myself. I allowed myself the possibility of becoming an astronaut and a scientist. By taking a step back and focusing all of my effort on one math class at a time, I mastered my weaknesses and opened an entire world of possibilities. I recall the exhilaration of this thought! I remember thinking back at that time to my early childhood, when I was perhaps seven, of seeing a reflection of a lamp, switched on in the darkness, suddenly appear in a distant mirror in our farmhouse. I recalled wondering if the image took time to appear. I always loved natural history. Suddenly, it seemed like the world became a more beautiful, understandable place once I had the tools to understand.

I left the service, and, with the help of the G.I. Bill and the money I had saved during my enlistment, I obtained a bachelor's degree in electrical engineering at the University of Washington. Even then, the idea of a career in pure science was distant, and it seemed to me that the possibility of failure was too great. Ever the practical person, I wanted to make sure I could find a secure job and support a family. The experience I had gained working on F-16s, together with my coursework, allowed me to land a position with NASA as a Space Shuttle engineer at Kennedy Space Center. Luckily, the F-16 and Space Shuttle shared many of the same technological features. I remember sitting one night on a catwalk on Launch Pad 39A (picture - http://tinyurl.com/ksclp39), watching the waves roll into the beach and eating a forbidden lunch out of my tool bag - the Space Shuttle being readied for launch just behind me. I saw a beautiful conjunction of planets and, for a brief moment, could almost sense the great wheel of the ecliptic. I felt a euphoric vertigo - a sense that I could fall right into deep space from where I stood. I was again called toward my old dream. This time it didn't seem so distant.

At this time, I began training to complete my private pilot's license in earnest. Having already completed Rescue and Advanced Scuba certifications while in Okinawa, I wanted to build a set of credentials that would make me competitive for NASA's Astronaut Corps. I decided, at this point, that I would make a major change in my career path starting with a move to NASA's Goddard Space Flight Center. I made the 26-hour round trip between Kennedy Space Center and Goddard fourteen times, and, after thoroughly wallpapering the Center with my resume, landed a job as the lead power system engineer on the Rossi X-Ray Timing Explorer (RXTE) spacecraft. I simultaneously began night school at The Johns Hopkins University to obtain a Masters degree in electrical engineering – specializing in multivariable control systems. After the successful launch of RXTE and two other spacecraft I lead as power system engineer - the Wide Field Infrared Explorer and the Transition Region and Coronal Explorer - I began to again become interested in the scientific aspects of these missions.

In my spare time, I read deeply in popular literature such as Sky and Telescope magazine and had conversations with many NASA scientists including Jean Swank, the principal investigator of the RXTE Proportional Counter Array instrument. Working very long hours integrating and testing RXTE prior to launch left me little time to do my schoolwork, but I managed to finish my degree in 1995 with ten graduate classes and a rather shoddy thesis in Quantitative Feedback Theory – an approach to multivariable control systems who's time never came.

Shortly after we launched RXTE, I began attempting to design my own telescopes. I read all of the books I could find at the library on telescope making and optics, but, being naturally cross-grained and in the habit of exploring paths seemingly at odds with my goals, decided against simply grinding mirrors using established principals or building anything that had been built by anyone else. With my friend, Peter Chen, working in the dust-filled dome of a decommissioned optical observatory on Goddard, we tried to find a way to build ultra-light mirrors out of composite materials in hopes of placing an observatory on the Moon using a Pegasus booster rocket launched from the wing of NASA's repurposed B52 bomber. Our mirrors, constructed by a process of placing multiple layers of composite and epoxy over a convex mandrel, were fabulously light. Peter's young daughter could easily hold one that was a full meter across – it weighing no more than a month-old puppy.

There were seemingly insurmountable 'print through' problems, however, that left the signature of the composite fabrics in the reflected light. (Interestingly, as I write this I can see several possible solutions including simply removing, or 'calibrating out,' those spatial frequencies using Fourier de-convolution in image post processing.) We also attempted to form meniscus mirrors by spinning liquid glue on a phonographic turntable, 'borrowed' from Peter's daughter, only to find standing waves frozen into the finished mirrors due to small, persistent vibrations from the cheap turntable bearings. Building on this 'result,' I conceptualized a Moon observatory in which the needed mirrors (the heaviest part of most telescopes) would be made there on the Moon, *in situ*, using lunar material heated to a liquid by solar power and robotically spun into an array of meter-class mirrors. The light gathered by these small telescopes could then be combined interferometrically over long, cold, naturally evacuated baselines – just like Young's double-slit experiment (video - http://tinyurl.com/h5qoakv) writ large – to provide staggering spatial resolution. I recall conceiving of a landing 'system' consisting of a flabby balloon that would cushion the robotic, self-replicating observatory as it impacted the Moon's surface and simply dissipated the energy of landing by rolling and bouncing. A retrorocket system would have been far too massive and complex for our approach to a Moon observatory 'on the cheap.' While these ideas never bore fruit, I still think they have some merit and could easily be adapted to other planets or moons. The planet Mercury, for example, would be ideal as it is tidally locked with the Sun, with one hemisphere permanently facing the dark of space. Such an observatory located near the terminator on Mercury - the line between the lighted and unlighted hemispheres of this tidally locked planet - would have almost unlimited solar power to reproduce itself using local material.

At about the time Peter and I were conducing these experiments, I met several design engineers at work on electro-optical mechanisms and science instruments. I began to 'hang out' with them on my off hours and eventually decided to change my specialty to the design and development of science instruments as an electromechanical engineer. This seemed like a natural step for me – and one step closer to the science that I found so interesting. I loved technical detail and positively wallowed in the minute specifications, capabilities and characteristics of electronics and optical parts – an absorbing activity that engineers often refer to as 'spec's-manship.'

Design was purest joy for me! Every morning I would come in early and read technical literature for about an hour – books on electrical grounding, remote sensing, spacecraft attitude controls - to 'up my game.' I was at that time assigned work on a prototype instrument for what was then named the Next Generation Space Telescope (NGST). The instrument was a Fabry-Perot interferometer (FPI) to study distant stars and galaxies by reflecting the incoming light between two half-silvered mirrors. My colleague and I were to design the instrument such that the mirrors could be kept parallel to one another to within a very small fraction of a wavelength of the science beam light. Because this radiation was in the near infrared, the FPI had to operate at cryogenic temperatures, greatly complicating our task. Upon determining that voice coil actuators would dissipate too much heat in the instrument, we machined mechanical amplifiers to extend the range of motion of ceramic piezo crystals that change dimensionally when a high voltage is applied. I designed a capacitive sensor system that provided sub-nanometer positional knowledge and a multivariable control system that decoded the signal from the three capacitive sensors, orthogonalized it and provided drive signals to the actuators to maintain parallelism. While this prototype was quite successful, the NGST science definition team decided against the inclusion of such an instrument on what is now called the James Webb Space Telescope (http://www.jwst.nasa.gov).

After several years of working on science instruments, I was chosen to be an engineer on the Wilkinson Microwave Anisotropy Probe (WMAP - http://map.gsfc.nasa.gov) – the mission that proved the existence of Dark Energy and permitted a very precise determination of the age of the Universe. In my spare time, I calculated that, unless the instrument's Sun shields deployed to within a very small margin of perfectly co-planar, the whole spacecraft would be 'spun up' by the momentum imparted by photons from the Sun, and would, consequently, expend all of its propellant years before the prime mission was to terminate. I showed my crude calculations to the Principal Investigator of the mission, Dr. Charles Bennett, who suggested that I return immediately to graduate school – this time to pursue a dual degree in physics and astronomy. I acted on this immediately.

Going back to school at The Johns Hopkins University mid-career, in middle age, was hideously challenging for me. Once again, I had to start over – re-teaching myself math from my old books, taking undergraduate classes in quantum physics, theoretical mechanics and statistical thermodynamics – subjects I knew almost nothing about. It was an amazing struggle! I recall taking radiative astrophysics – the most challenging class I have ever attended from JHU's toughest physics instructor – Professor Julian Krolik. Seated at the very front of the room, I would place a small recorder near Prof. Krolik and take notes on the left half of each of my notebook's pages. I would then, immediately go to the library and retake the same class from my tapes – updating my notes with anything I had missed. I remember, lying in my sleeping bag on my office floor at JHU, thinking bitterly that I was simply not smart enough – an impostor that should never have been there studying astrophysics, wishing I had never left the farm of my youth. I had no idea that I had suffered severe ADD my entire life – a condition I am now being treated for. The brutal mechanisms I contrived to survive college and graduate school, through three degrees in engineering and two in physics and astronomy, nearly crushed the very spirit out of me. After ten brutal years, I delivered my doctoral dissertation – a dense, forbidding, esoteric tome on long-baseline stellar interferometry including the mathematical foundations for and first science with the Keck Interferometer Nuller (KIN - picture: http://tinyurl.com/keckobs) that combined the light from the World's two largest optical telescopes on the peak of Mauna Kea in Hawaii. KIN was NASA's pathfinder for an orbiting observatory, the Terrestrial Planet Finder Interferometer (TPFI), a six billion dollar project that was to search distant worlds for signs of life.

Within weeks of my graduation from JHU, NASA stopped all work on TPFI – the mission for which I was being groomed. It was time to start over again.

In a disoriented, suicidal haze, I applied to NASA's Astronaut Corps. With five degrees, Space Shuttle launch experience, military and skydiving experience, scuba and piloting certifications, and near-perfect vision I thought I had a chance. I remember waiting for the letter, hoping for some way out – some way to amend this final impasse. I waited for months for what was to be a perfunctory, unsigned rejection note – a note I still have, tucked away in a drawer. I had, it turns out, antibodies against my own thyroid gland in my blood – a condition easily and completely correctable with inexpensive medication – a tiny pill I now take once a day.

I actually don't remember what it felt like, the crushing finality of that rejection. I vaguely recall angry, futile arguments with NASA's flight surgeon at Johnson Space Center. I remember demonstrating the absurdity of his assertion that radiation would denature any medication they sent up into space with me – all to no avail. Looking back, I feel oddly liberated. I recall always thinking of every sickness, every injury as being a potential 'show stopper' for my application to the Astronaut Corps. I remember desperately doing eye exercises to keep my vision acute and sweating at JHU's gymnasium between marathon study sessions. All of it was gone. The dream that I had worked towards for twenty years was irretrievably gone in one moment.

In an attempt to resurrect a space borne interferometer, my colleagues and I designed the Fourier-Kelvin Stellar Interferometer. This was to be a greatly simplified version of the Terrestrial Planet Finder Interferometer - a pathfinder like the Keck Nulling Interferometer, but in space. We proposed this to NASA and did get some funds to build an optical testbed. Spending long hours in the laboratory, I was able to demonstrate broad band nulling of a simulated science beam, but the mission never was accepted for full funding. There is a great deal of resistance in the astronomical community to all sparse aperture techniques as it requires one to think in Fourier Space when integrating the light from separated telescopes. The community advocating for this technique was simply too small, the technique too esoteric and technically challenging to sell to the larger astronomical community. At some point in the future, I am quite certain that we will need to exploit this technique to assay the characteristics of distant Earth-like planets. For now, advanced coronography - nulling the light of the central star by obscuring it optically - is under active development. It is more accessible for most astronomers but will never achieve the spatial resolution that can be provided by a long-baseline interferometer.

At this time, I was fortunate to become involved in a new project using NASA's Deep Impact probe to study exoplanets. The main purpose of this spacecraft was to fly to a comet nucleus and fire a 300 kg penetrator probe at it. Powerful telescopes on the main 'flyby' spacecraft would then monitor the impact of the probe to determine the molecular abundances in the comet using spectroscopy. Comets are thought to consist of primordial matter from the formation of the Solar System so understanding their chemical makeup is of great importance to planet formation theory. The primary mission was reasonably successful, but was greatly hindered by a defect in one of the primary telescopes on the flyby spacecraft. This 'High Resolution Instrument' (HRI) was calibrated in a cryogenic chamber that simulated the space environment prior to launch to assure that the focus mechanism would have sufficient dynamic range to respond to the dimensional change in the length of the telescope due to exposure to the very cold space environment. Unfortunately, the optically perfect flat mirror that was used to conduct the calibration became warped in this environment for this same reason resulting in it becoming a 'powered' optic. The focus mechanism was calibrated without knowledge of this powered optic being in the light path. Consequently, the telescope was launched with an uncorrectable defocus anomaly.

Our idea was to capitalize on this defocus anomaly to conduct transit photometry of exoplanets. We realized that if a telescope has high spatial resolution and a star with a transiting planet is in precise focus, this light will fall onto a single pixel. Consequently, slight spacecraft pointing shifts coupled with the non-uniform inter and intra-pixel sensitivity of each pixel will result in a systematic time and temperature variable signal to be imposed onto the astronomical signal. If the light could be integrated across a number of pixels, this effect could be greatly reduced. The best way to accomplish this is to defocus the telescope!

Using the permanently defocused HRI to advantage, we studied several transiting planets and conducted photometry leading to several important findings. In particular, earlier analysis of data for several exoplanets we observed suggested the possibility of companion masses in their stellar systems. Exhaustive observation and transit timing analysis of these planets: HAT-P-4, TrES-3, TrES-2, WASP-3, and HAT-P-7 resulted in new constraints on any other planetary companions in these systems.

My particular part in these studies was to construct point spread functions (PSF) for the instrument - to be used to isolate the light from the star and planet. A PSF is the systematic behavior of how the entire telescopic system responds to a point of light. In the case of the HRI telescope, this was a doughnut-shaped blob due to the defocus anomaly. Unfortunately, this anomaly was also strongly chromatic - the shape of the PSF changed with the color of the star and the color of the filter used! I realized that because color tracked to PSF shape, there might be a way to use this to distinguish the light from the star with the planetary companion from other stars in the near vicinity using color. By carefully constructing a PSF for the precise color of the planet hosting star and deconvolving the image in Fourier space with that PSF, the only star in the image that became point-like due to the deconvolution was the one of that same color - the planet host! My experience in Fourier math gained from my work in long-baseline stellar interferometry had shown me a way to capitalize on the chromatic defocus anomaly in the HRI instrument!

Fortunately, my work on the PSFs for the HRI instrument was noticed by a gravitational microlensing group at Notre Dame University lead by David Bennett. With Dr. Bennett, myself and several other scientists, we proposed to use these same techniques to detect *microlensing* exoplanets in observations of the Galactic bulge while the Deep Impact spacecraft was cruising to its next cometary target.

Microlensing requires some explanation. Gravitational microlensing is a relatively new technique conceived by Albert Einstein but not technologically possible until fairly recently. At the most fundamental level, microlensing may be understood by thinking of light as rays. If spacetime is flat, rays of light will travel radially from their source in perfectly straight lines. If we introduce any mass, spacetime becomes warped around it. If a ray of light passes through the gravitational potential well formed by the mass, its path will be deflected towards the mass. If some massive object came into alignment between an observer and a distant light source, the observer would see a brightening of the distant light source because rays of light that were traveling radially from the source were deflected inwards by the mass. The distant light source would be magnified by the mass - as if a classic lens was passed in front of it. To observe the actual images of the distant source created by the lens would require a telescope with an aperture the size of Earth's orbit. Luckily, the flux from the source is conserved by the warping of spacetime, so this magnification is manifested by a brightening of the source star - a brightening that may approach several hundred times the ground-level brightness of the source. This beautiful phenomenon may in fact be understood and modeled by a simple lens equation at its most basic level!

Because the alignment of the source, the lensing mass and observer must be very precise in order to create enough magnification to be detected by current instrumentation, detection of a gravitational microlensing event is exceedingly unlikely. For this reason, the technique is applied to stars in the Galactic bulge. The bulge is a very dense, very old roughly football-shaped spheroid consisting of billions of stars. These stars orbit about the super-massive black hole at the center of our Galaxy. Most of the lensing bodies detected so far are Galactic disk stars some of which have planets in orbit about them.

If you have been reading along carefully, you will note that the lensing mass does not need to emit any light at all. The lens is detected only by the deflection of light from the source due to the lenses warping of spacetime. Consequently, this technique is capable of detecting lensing bodies that can be detected by no other method because they are too dim - or emit no light at all! These could include compact massive objects such as black holes and neutron stars or even planets that are not gravitationally bound to any star! Such free floating planets have long been predicted by planet formation theory but have never been detected because such a planet would emit very little light - only a dim glow from the heat of its formation.

What can gravitational microlensing tell us about planets? Because this technique is sensitive to planets in distant orbits from their host stars or even unbound planets, it can provide information that no other technique can. Other methods to find and characterized exoplanets are transits, radial velocity detection, and direct imaging. The first two of these reach peak sensitivity for very massive planets close to their host stars. The last is in its infancy and is only sensitive enough to detect planets in orbit about nearby suns. Of all of the available exoplanet detection techniques, gravitational microlensing is the only one capable of detecting a large enough number of exoplanets in distant or unbound orbits to permit the calibration of planet formation theory. Current formation theory tells us very little about the frequency and characteristics of close-in exoplanets. These theories do, however make predictions about planets in distant orbits or unbound planets. Consequently, the only way to determine which theory or model is correct is by comparison between the predictions made by each model and a statistically robust observed sample of such planets - something only gravitational lensing can provide.

But why is this important? Why should we be interested in the formation of planets far away from their host star? Lets start closer to home! Current planet formation theory tells us that Earth, if she had formed from dust and planetesimals where she is now - at her current distance from the Sun - should be bone dry. There should be no water, no oxygen, no volatiles of any kind. Where did our oceans come from? Why do we have an atmosphere? Why do plants and animals - and humans - exist? Even the most optimistic accounting of cometary impacts on Earth have grave difficulty in accounting for Earth's large oceans and atmosphere. How did this happen? How are we to look for other life bearing planets in the Galaxy if our best theories tell us that planets near enough to their host star to have liquid water at their surface should have little or no water! Current theory suggests that simply having a planet of approximately the correct mass at approximately the correct distance from its star to have liquid water at its surface is no guarantee at all that the planet is Earth-like, that is able to bear life or, perhaps, life bearing. On the contrary, current planet formation theory - until now completely uncalibrated by a robust sample of long orbit planets - gives us almost no guidance at all as to where to look for an Earth-like planet. Once we have advanced our planet formation theories to the point that we can account for Earth's oceans, then, *and only then*, will we know where to point our telescopes to look for a Sister Earth. It is entirely possible - even likely - that Earth did not form where she is now. She may have been a distant planet or a moon of one of the gas-giant planets in the outer Solar System that migrated in to her present clement orbit. This is a hypothesis that I have recently put forward. Such a plausible hypothesis may one day be validated by the observations of the warping of spacetime by distant stars and planets!

My lab is in Building 34 of NASA, Goddard Space Flight Center in Maryland. This one building, home to NASA's Laboratory for Exoplanets and Stellar Astrophysics, the Laboratory for Exobiology, the Laboratory for Planetary Chemistry, The Laboratory for Experimental

Cosmology and other related groups, represents the greatest concentration of astrophysicists in the world. Just outside of my office, three floors down, there is a large, deep depression in the ground, about 20 feet from the base of our building. There is no obvious reason for this structure. There is no need for a deep ditch like this, no drainage issue that it addresses. This deep depression with a grating at the bottom does serve a particular purpose. Building 34 is quite new, designed and built over the past five years. The winning architect thoughtfully designed this depression as a *tank trap*. Placed at the most vulnerable point in the building, it is designed to stop a truck bomb from killing the occupants.

Why?

Very soon, science will discover life somewhere off planet. I expect this to happen within the next twenty years or so. Through all of human history, we have been alone in the Universe. We live during an unprecedented confluence of scientific thought and technological advancement that will yield compelling evidence for alien life and this will happen within the next few decades. This will occur during a period in human history when religious fundamentalism and the denial of the scientific method are both troublingly on the rise.

The importance of this time cannot be understated. It will be an epochal turning point . Prior to this generation - the generation alive today - humankind was utterly alone in the Universe. At one point in time in the very near future - and forever after - we will not be alone. It will be, in a very real sense, childhood's end for our species. What will humankind do when confronted with this new, disorienting truth? The architect of Building 34, at least, anticipated civil unrest, perhaps violence. I am sorry to say that I do, too.

A Rolling Stone Gathers Moss (even from Ötzi the Iceman's Innards)

James H. Dickson

Jim Dickson was educated at the Universities of Glasgow and Cambridge. In the 1960s he was a Research Fellow of Clare College and Senior Assistant in Quaternary Research in the Botany School in Cambridge. The author/co-author of six books and some 150 (and counting) scientific papers and reports, he retired from Glasgow University in 2002, latterly having been Professor of Archaeobotany and Plant Systematics. In 2012 the Dickson Laboratory of the York Archaeological Trust in the Kelvin Campus, Science Park, Glasgow, was named in his honour, this honour being shared with his late wife Camilla, an eminent archaeobotanist. Jim is a Fellow of the Royal Society of Edinburgh and a Neill medallist of that Society.

In 1991 the discovery of Ötzi the Iceman was one of the greatest in all archaeology and is likely to remain so. Utterly unexpected, that such a happening very high in the Alps could come about had occurred to nobody. Here was a well preserved if shriveled body of a man, with clothes and equipment, from no less than 5,200-5,300 years ago. A great deal about Ötzi is unique in archaeology, certainly from such a remote period; for instance there was an entire copper-headed axe, a fire-making kit and a complete set of archery equipment and lots more. On first hearing about it I was entranced. Would there be mosses found with him - just a few perhaps or maybe up to twenty species - few mosses can grow at such a high altitude (3210m a.s.l. or about 10,500 feet). When I was finished studying them there were no less than 80 different in total (including the related liverworts) and very informative many of them are. The fragments were in him, on him and around him in the icy hollow where the frozen mummy had melted out of the ice. What do they tell us about Ötzi? How did my involvement in this totally absorbing, celebrated, very cold case investigation come about (and not just because my middle name happens to be Holms, yes, no *e*)?

Early Years in Glasgow: Birds, Plants and First Discovery

Birds were my first natural history passion. That was in the mid to late 1940s. I had Edmund Sander's *A Bird Book for the Pocket* and later James Fisher's two paperbacks on *Bird Recognition* and finally his *Watching Birds*. I still have these four ornithological books and they bear the marks of much use long ago. As a youngster some birds seemed almost mythical to me. I would never see them: Golden Orioles, Red-necked Phalaropes, Great Northern Divers and Hoopoes. Family holidays to the Isle of Bute or the Isle of Arran, both in the Firth of Clyde not far west of Glasgow, could never produce them; Oyster Catchers and Gannets were all very well but not in my eyes to be compared with these well out-of-reach creatures.

James Fisher's books were published by the appropriately named Pelican Books, an offshoot of the economically priced Penguin Books, all paperbacks. When a teenager and young man such books were very important for my education. I still have eleven Pelicans on natural history, twelve on other sciences, nine on archaeology, two on religion and one on mountaineering. No doubt there were others which have disappeared over the decades, though I am very possessive, perhaps overly so, of my library.

I clearly remember when on boyhood holidays being conscious not just of birds but wild flowers too. When I was taught the structure of flowers at secondary school, Bellahouston Academy in Glasgow, I took it from there and hunted plants seriously. After all, plants are stationary but birds are not (and in any case I had no binoculars and no friendly birder/twitcher to help). I was and will always be fascinated by the enormous variety of nature and my great aim was to learn the names of every plant and fungus (in those days fungi were treated as plants; we know better now). Where did this drive come from? Really I do not know but my suspicion is more from my mother (whom I remember clearly only as slowly dying, bed-ridden with MS) than my father (with whom I cannot recall having had any even remotely intellectual conversation).

So first there were birds, with plants a very close second and then archaeology was not so very far behind. When a boy I was taken if need be by a loving aunt to be treated by homeopathy. In the doctor's waiting room there were usually copies of the large-paged *Illustrated London News*. That now long defunct magazine always had a full page of archaeology which I very much enjoyed reading.

With the great thirst for knowledge of biodiversity stronger than ever in 1955 I went to the University of Glasgow with the already firm intention of reading botany. Though very well-meaning the lecturers as a whole I found not very inspiring and in one or two cases disappointingly pedestrian. The one who influenced me most was Alan Crundwell, an authority on mosses and liverworts, collectively called bryophytes. He was a very good botanist in the field and therefore a soul mate, as far as an undergraduate and lecturer could be such. I collected a moss that puzzled me from Ben Vorlich, a mountain near Loch Lomond, north-west of Glasgow. I knew what it was approximately but not with confidence the very particular species. Alan Crundwell told me it was Bright Silkmoss, till then unrecognised in Britain and that, of course, pleased me greatly; it has since been shown to be widespread in Britain. Mentioning me, he published the discovery but I was not a co-author of the two page paper.[1] No matter, I had taken up bryology not just keenly but successfully.

Over the next sixty years, apart from Britain and Ireland, I have collected mosses and liverworts in Denmark, Norway, Sweden, Finland, Germany, Austria, Italy, Switzerland, France, Andorra, Poland, China, Nova Scotia, Newfoundland, British Columbia, Yukon, Northwest Territories and Alaska as well as the Canary Islands off the north-west coast of Africa, and on Ascension Island and St Helena. I collected even on the Tristan da Cunha islands very remote in the South Atlantic Ocean. That was when I was botanist on the Royal Society expedition to study the volcanic eruption of 1961 and what an experience that was for a young man still not twenty five to visit such an archipelago; my task was to study the effects of the eruption on the vegetation but I gathered bryophytes whenever I could. These have been studied in detail only very recently and it turns out that I had collected a species of liverwort new to science and another liverwort of a genus new to science. Rather than use the botanical

Latin if I can concoct English names, the first is Tongue-shaped Featherwort and the second is the Deceptive Featherwort.[2]

So this rolling stone really can claim to have gathered much moss very widely and not just living mosses but sub-fossil ones too, as I now relate.

Cambridge in the 1960s: Ice Age Mosses and Bogmoss as a Bronze Age Wound Dressing?

Leaving Glasgow in late 1959, with Alan Crundwell's backing, I found myself very well placed when I did graduate research in the Sub-department of Quaternary Research of the Botany School at the University of Cambridge. This led eventually to my book "*Bryophytes of the Pleistocene*"[3] - not the catchiest of titles I readily admit; it would have been better called "*Ice Age Mosses.*" My principal mentor in Cambridge was Harry Godwin, later to become Professor Sir Harry, who was the great British pioneer of botanical investigations of the Quaternary (the Ice Ages, roughly the last two million years or so of the Earth's history). He and his graduate students had been concerned with pollen grains, spores, seeds, leaves, wood and the like, that is to say the sub-fossil remains of flowering plants, conifers and ferns. My task was to see how mosses fitted into the results obtained from the other kinds of plants - Just my cup of tea! Plant hunting in the recent past; what could be better? The remains of mosses preserved in the glacial and interglacial layers were in many cases those species already familiar to me as part of the present flora of Scotland and England but even more engagingly there were species, though not very many, that no longer grow in Britain.

One of the first bits of work in Cambridge I did was the identification of sub-fossil mosses from the late-glacial layers (over 13,000 years old) at Loch Droma in the far north-west of Scotland.[4] The outstanding moss was Northern Haircap never before recognised from the British Quaternary. It is a good indicator of the past environment because it grows only on acidic slopes where the beds of snow lie late into summer. Later I found this moss in late-glacial layers at low altitude in Cornwall, south-west England; at present in Britain it grows only at high altitudes in the Scottish Highlands. There were two authors of the paper resulting from the Loch Droma studies, Harry Godwin being one and he quoted whole paragraphs of my first year report including a purple passage about Northern Haircap but I was not the other author. It's very changed days. Now it is utterly routine to see papers with the very many authors such as the forty of a recent paper on Ötzi the Iceman.[5]

From the late 1940s Harry Godwin had had dealings with Willard Frank Libby, the Nobel prize-winning inventor of radiocarbon dating. By the mid-1950s Harry Godwin had had constructed by Eric Willis one of the first radiocarbon dating apparatuses in a British University. Consequently archaeologists were very keen to send samples for dating. Together as young men in the 1930s Harry Godwin and Graham Clark, later Professor Sir, Disney Professor of Archaeology in Cambridge, had investigated the peat at Shippea Hill near Ely in the Fenlands of East Anglia.[6] There Bronze, Neolithic and Mesolithic debris was stratified in the peat. This was one early "environmental archaeology" investigations carried out in Europe - of course that was long before that term was invented. Just as I was arriving in Cambridge, the two academics decided to re-excavate to obtain samples for radiocarbon dating. Harry Godwin asked me to be the "environmental archaeologist" and so off I went to get suitable samples for botanical analyses as well as radiocarbon dating from the re-excavation in 1960. Here was I assisting one of the most famous archaeologists in the world. This was pure mind-riveting fun and Graham Clark took a kindly if somewhat distant interest in me thereafter. The resulting dates helped in pushing the onset of the Neolithic in Britain back many hundreds of years to around five thousand years ago, or about six thousand years ago if account is taken of dendrochronological calibration.

In the early 1960s there arrived a sample for dating taken from a Bronze Age cist grave in Fife, Scotland. The body of a young man had been buried with a dagger and a beaker and there was also "black, crumbly matter" over the floor of the cist but thickest near the skeleton's chest. Harry Godwin's technical assistant, Camilla Dickson (née Lambert), examined the matter microscopically and produced an unexpected pollen count; there were high values of Linden and Meadowsweet, indicative perhaps of funereal flowers and/or honey or mead from some three thousand or more years ago. The significance of Meadowsweet pollen found in archaeological contexts is an ongoing topic in Scotland, Wales and on continental Europe as well. Is it indicative of eating of honey, consumption of alcohol or bunches of sweet-smelling flowers? [7]

Camilla handed over to me some pieces of mosses she had extracted from the black crumbly matter. Apart from Glittering Woodmoss, there was a Bogmoss (Blunt-leaved Bogmoss to be precise) which made me very intrigued. Why? Well, right down to very recent years Bogmosses are well known as wound dressings; was that why the Bogmoss was in the grave? Had the young man been wounded (perhaps fatally) and his wounds staunched with Bogmoss - a very engaging, plausible thought but is it true? One upshot of this collaboration was that Camilla and I were married soon after.

Back in Glasgow University: Roman Toilet Habits and what the Romans did not do to Scotland

Back in Glasgow in 1970 with Camilla I set up the first lab devoted to archaeobotany and Quaternary botanical studies in Scotland. One of the first major works we started was the archaeobotany of the Roman fort at Bearsden to the north-west of Glasgow. In the 140s AD, the Roman army built a wall, the Antonine Wall, across Scotland from the Firth of Forth to the Firth of Clyde. Among the many forts and fortlets along the wall the one at Bearsden was rather small and very briefly occupied. There is a suggestion that it was garrisoned by "*Cohors IV Gallorum quingenaria equitata.*" (What did these cavalrymen from Gaul think of the cool, wet and windy climate of the west of Scotland - Delightful or otherwise?). The products of the latrine accumulated in one of the defensive ditches and little did the troops know when relieving themselves how very interesting the investigation of that layer would prove to be nearly 2,000 years later. (Still smelly particularly if hot water was applied, I may add.) We found Figs, Coriander, Dill, Opium Poppy, Lentils, Linseed and many other coarse remains of edible plants as well as eggs of the intestinal parasite Whipworm. There were pieces of mosses too in the sewage impregnated layer including Flat Neckera, Glittering

Woodmoss, Neat Feathermoss, Slender Mousetail Moss and Springy Turfmoss.[8]

Why do I mention these particular mosses? It is because they are large species suitable for small scale domestic purposes such as hygienic wipes - which is just a way of saying they could well have been the equivalent of toilet paper and very suitable they are. In the literature concerning the Romans it is often stated that they used a sponge on a stick. However, this is based on very sparse Roman writings. It seems probable that at Bearsden the troops used these large mosses, readily available in the vicinity. Perhaps also mosses were used elsewhere in the empire where such suitable mosses could be gathered but not in places like the deserts of the Middle East and North Africa. There is incontrovertible archaeological evidence of mosses having been used as toilet paper at various places and times through prehistory and history, as for instance by the Vikings in Dublin, Ireland.

The Romans use of locally gathered mosses would hardly have produced any great visible effect in Scotland. However, what did the Romans see when they invaded Scotland in the first and second centuries after Christ? Were they hindered by enormous stretches of dense forest or not? The first learned Scot to think about that 300 years ago thought it was the Romans who removed the trees to produce great areas of openness; they and only they could have had the capability. Such an opinion prevailed for a long time, far longer than it should. Now we know better. The pollen diagrams from bogs and lake muds we produced to investigate that matter in west-central Scotland show that the forests had been extensively cleared by the Iron Age people before the Romans came. There is a marked rise in the proportion of grass and weed pollen with a coinciding fall of pollen of trees. In some places it was hundreds of years before while in others only just before but nonetheless before. Elsewhere in Scotland the first big forest clearances began earlier, for instance, in Neolithic times on the Orkney archipelago and Bronze Age times on the Isle of Arran.[9]

If only briefly and intermittently in forays to the north of the Wall, the Romans would have set eyes on Loch Lomond, the lake already mentioned above. It is the largest area of fresh water in Britain and most of its volume is below sea-level. Because of the interplay of rise and fall of the land and sea-level, twice the loch has been joined to the sea in the last 12,000 years or so, firstly at the end of late-glacial times and then again some few thousands of years later; the second incursion had been suspected but not proven. We produced pollen and diatom analyses from the mud cored from the south basin and this gave the proof; deep in the mud there is a layer of diatoms indicating full sea water salinity with fresh water ones above and below. The second marine incursion lasted some 1,500 years or so. This is the only example of such a history for any lake in Britain.[10]

Mosses had nothing to do with my leadership of the five man Trades House of Glasgow expedition in 1987 to Papua New Guinea; we collected orchids, begonias and ferns to be grown at Glasgow Botanic Gardens and shown at the Glasgow Garden Festival in 1988, a very popular event. Nor were mosses included in two books on the urban and rural plants of the Glasgow area in 1991 and 2000 - the fulfilment of an ambition held since my teenage years.[11] However, mosses, as already mentioned, had a very great deal to do with the next topic.

Ötzi the Iceman: Moss Fragments in every Gut Sample

Now I tell of what is perhaps the great highlight of my career. For me it is an utterly satisfying combination of my interests in mosses and in archaeology. Immediately after the discovery, the Iceman, a well-preserved natural mummy found in the Alps on the border of Austria and Italy by hikers in 1991, was world famous and rightly he has remained so. Often he is called the Tyrolean Iceman or Ötzi (after the Ötztal, valley of the river Ötz, and pronounced to rhyme with tootsie).[12] What had he been doing so high in the Alps? Which direction had he come from? Was it the north or the south? What was his way of life? Why was he carrying a useless set of archery gear? Why was he fatally shot in the back with an arrow? Why did the killer leave the copper-headed axe which many archaeologists regard as having been a valuable, status symbol? The intriguing questions go on and on. The mosses have a great bearing on some of these matters. The archaeobotanical studies have proved to be highly productive ones among all the very many investigations which have revealed so much of the man's way of life, his environment and the violent death in his mid-forties.[12]

In April 1994 samples of the many mosses found with the Iceman arrived in my lab from Innsbruck University, Austria, where the mummy was at that time. With high expectations I opened the package and saw lots of glass vials, some much fuller than others. I took the stopper out of the one which seemed totally crammed. With tweezers I began gently to pull the mass of moss out. Before it was completely out I knew what it was. Flat Neckera, yes, Flat Neckera! Eureka! Why the great delight? The Iceman was found at 3,210m a.s.l. (about 10,500 feet), well above the tree line. A woodland moss, Flat Neckera, does not grow above about 1,700m a.s.l. in that area and so how come it was with the Iceman? The great likelihood was that he had deliberately carried that moss (and others not considered here), for whatever purpose, up the mountain. Additionally, Flat Neckera is a substantial part of the archaeobotanical evidence that he had been going from south to north on his last journey.[13]

The Iceman mosses had been sent to me by Professor Klaus Oeggl, a well-known archaeobotanist. Klaus and his graduate students have done outstanding work on the Iceman. A particularly elegant investigation concerns the contents of the Iceman's alimentary tract; this resulted in a multi-author paper with Klaus in the lead: *"The reconstruction of the last itinerary of Ötzi, the Neolithic Iceman, by pollen analyses from sequentially sampled gut contents."*[14] In the last 36 hours or so of the Iceman's life he had been high in the mountains then low down perhaps as low as about 1,000m a.s.l. and finally very high up indeed where the mummy was found - in the zone of perennial snow and ice where plants can grow only very sparsely. These deductions are unique in archaeological science. How were they made? The distinctive walls of pollen grains are not digested in the human gut and so extracted and spread on microscope slides they can be identified and counted. The different altitudinal zones of vegetation such as deciduous forest, coniferous forest and alpine grassland have different pollen signatures. So from the faeces in the rectum we can infer high altitude, from the colon content low altitude, and from the ileum and stomach high altitude again.

The most crucial pollen grains from the colon are those of the small tree of low altitudes called Hop-hornbeam, well known to grow to the south of where Ötzi was found. Another important recovery from the colon was a tiny fragment of a Bogmoss, most certainly a Bogmoss and probably Imbricate Bogmoss, a species which grows at low altitudes in Europe and I stress never high in mountains. When samples were obtained from the stomach, just a few years ago, there was another very small fragment of a Bogmoss, again certainly a Bogmoss but I cannot be sure of the particular species. How did these fragments of Bogmosses get into in the Iceman's colon and stomach? I consider that they had been part of a wound dressing fragments of which he had swallowed accidentally. The Iceman's right palm had been badly cut shortly before he was killed, perhaps a day or two earlier; we know that because the wound had already begun to heal. He used some Bogmoss to staunch the wound and this meant that little bits of Bogmoss stuck to his bloody fingers. So when he ate or drank some of these little pieces got into his innards.[15] As tiny fragments of Flat Neckera has been found in all the gut samples except that from the stomach, I think that this moss had been used as wrapping for the dried meat Ötzi carried and so the fragments got into his innards.

Though by far the oldest, Ötzi is not the only mummy to have had fragments of Bogmosses in the alimentary tracts. There are at least two other frozen mummies; one is that of one of the women who lived about 1475 A.D. in western Greenland and the other is that of Long Ago Person Found, an indigenous young man who lived a few hundred years ago and melted out of a glacier in northernmost British Columbia.[16] Why were the Bogmosses there? Here is a topic needing more investigation as other ancient ice mummies are found.

Because of some deaths of investigators of Ötzi soon after the discovery, journalists invented the preposterous Curse of Ötzi, like the more famous, but equally silly, Curse of Tutankamun. As I write this I am 76 and Ötzi has been my benefactor for over twenty years. In 2004 I received an email from a French student, a lady of mature years who was writing an essay on Ötzi. Less than three years later I married Geneviève Lécrivain, a retired school teacher and devotee of Palaeolithic flints. Together we much enjoy visiting prehistoric and historic sites across

Europe and the Mediterranean countries, particularly in France. Of course, I take the chance to hunt for plants too - it's in the blood or it's in my DNA as they say nowadays.

Audio: James Dickson discusses his Ötzi discoveries in this WNYC Podcast - Radiolab - An Ice-Cold Case - www.radiolab.org/story/ice-cold-case/

Video: Lecture at the Goddard Scientific Colloquium - NASA - James Dickson - Why Ancient Glacier Mummies Are So Important: The Intestinal Contents are Encoded Maps and Diaries - tinyurl.com/dicksongoddard

Video: Lecture - Bogmosses and Ancient Ice Mummies by James Dickson at the Beaty Biodiversity Museum, University of British Columbia, Vancouver, Canada, in 2012 - http://tinyurl.com/bogmossesmummies

Keep right on to the End of the Road

Finally, what of the birds I took as impossible to be seen when I was a young boy? My choice of career was rewarding in more ways than one. Because of my scientific and archaeological work I have been able to visit many parts of the world. This did more than satisfy my long standing wanderlust. In the summer of 1969 I heard the mysterious calls of Great Northern Divers in the Mackenzie Delta area of Arctic Canada. In 1973 I saw Hoopoes on the Mediterranean island of Majorca. In 2002 I saw Red-necked Phalaropes doing their synchronised swimming reels at Yakutat, southeast Alaska. In 2013 in Haute-Saône, north-eastern France, as I sat in Geneviève's garden, I heard the distinctively melodic calls of Golden Orioles. At last all the fabulous birds are real.

Right now I am identifying Neolithic mosses from Zurich, Switzerland; there is no Flat Neckera but lots of the related Crisp Neckera. Both of these species have been much used by prehistoric and later people for many purposes such as packing, wrapping and wiping as well as caulking boats (from the Bronze Age in England till as late as 1960 A.D. in France). The more we know about our ancestors like the incomparable Ötzi the better we understand ourselves. Technologically advanced as we are, we understand better how intelligent, observant, skilled and inventive our remote ancestors were. Kept frozen in the Iceman Museum in Bolzano, Italy, Ötzi will tell us more and more as more and more scientific techniques are invented.

The scientific natural historian and archaeologist in me cry out in unison "What a wonderful world!" While recognising that great change is inexorable, we must strive endlessly to keep it that way.

References

1. Crundwell, A.C. 1959. *Plagiothecium laetum* in Britain. *Transactions of the British Bryological Society* 3, 63-64.
2. Dickson, J.H.1965.The Biological Report of the Royal Society Expedition to Tristan da Cunha. General Introduction. Philosophical Transactions of the Royal Society of London B 249, 259-272. Váňa, J. and Engel, J. J. 2013. *The Liverworts and Hornworts of the Tristan da Cunha Group of Islands in the South Atlantic Ocean.* Memoirs of the New York Botanical Garden 105. New York Botanical Garden Press. The newly decribed liverworts are Adelanthus lingulatus and Deceptifrons plagiochiloides.
3. Dickson, J.H. 1973. *Bryophytes of the Pleistocene.* Cambridge University Press.
4. Kirk, W. and Godwin, H. 1963. A Late-glacial Site at Loch Droma, Ross and Cromarty. *Transactions of the Royal Society of Edinburgh* LXV, 225-249. The disussed moss is *Polytrichum sexangulare.*
5. Keller, A. et al. 2012. New Insights into the Tyrolean Iceman's origin and phenotype as inferred by whole-genome sequencing. Nat Commum 3, 698. Doi: 10.1038/ncomms1701.
6. Clark, J.G.D. and Godwin, H. 1962. The Neolithic in the Cambridgeshire Fens. Antiquity 36, 16-23. West, R.G. 1988. *Biographical Memoirs of the Royal Society* 34, 260-292. Fagan, B. 2001. Grahame Clark *An Intellectual Biography of an Archaeologist.* Westwood Press; Cambridge MA.
7. Dickson, J.H. 1978. Bronze Age Mead? *Antiquity* LII, 108-113. Dickson, Camilla and Dickson, James. 2000. *Plants and People in Ancient Scotland.* Tempus; Stroud. Linden is *Tilia* and Meadowsweet is *Filipendula ulmaria*. The mosses are *Hylocomium splendens and Sphagnum palustre.* Moe, D. and Oeggl, K. 2013. Palynological evidence of mead: a prehistoric drink dating back to the 3rd millennium B.C. *Vegetation History and Archaeobotany xxx.*
8. Breeze, D. and 26 others including Dickson, C and Dickson, J.H. 2015. The Roman Fort on the Antonine Wall at Bearsden. *Proceedings of the Antiquaries of Scotland* in press. The mosses are *Hylocomium splendens, Neckera complanata, Isothecium myosuroides, Scleropodium purum* and *Rhytidiadelphus squarrosus.* Hobson, B. 2009. *Latrinae et Foricae. Toilets in the Roman World.* Duckworth; London.
9. Boyd, W E 1984 Environmental change and Iron Age land management in the area of the Antonine Wall, central Scotland: a summary. *Glasgow Archaeological Journal* 11,

75-81. Dickson, J H 1992 Presidential address Scottish Woodlands; their ancient past and precarious present. Botanical Journal of Scotland 46, 155-165. Dickson, J H, Dickson, C, Boyd, W E, Newall, P J & Robinson, D E 1985. The Vegetation of Central Scotland in Roman Times. *Ecologia Mediteranea* XI, 1. Keatinge, T.H. and Dickson, J.H. 1979. Mid-Flandrian Changes in Vegetation on Mainland Orkney. *New Phytologist* 82, 585-612. Miller, J., Dickson, J. H. and Dixon, N. 1998 Unusual Food Plants from Oakbank Crannog, Loch Tay, Scottish Highlands: cloudberry, opium poppy and spelt wheat. *Antiquity* 72: 805-811. Ramsay, S 1996 Human impact on the vegetation around Walls Hill. Pages 59-63 in Alexander, D (ed) Prehistoric Renfrewshire Local History Forum. Ramsay, S. and Dickson, J. H. 1997 Vegetational History of Central Scotland. *Botanical Journal of Scotland* 49: 141-150. Robinson, D.E. and Dickson, J.H. 1988. Vegetational history and land use: a radiocarbon-dated pollen diagram from Machrie Moor, Arran, Scotland. *New Phytologist* 109, 223-251.

10. Dickson, J.H., Stewart, D.A., Thompson, R., Turner, G., Baxter, M.S., Drndarsky, N.D. and Rose, J. 1978. Palynology, palaeomagnetism and radiometric dating of Flandrian marine and freshwater sediments of Loch Lomond. *Nature* 274, 548-553.

11. Dickson, J.H. 1991. *Wild Plants of Glasgow*. The Mercat Press; Edinburgh. Dickson, J.H., Macpherson, P. and Watson, K. 2000. *The Changing Flora of Glasgow*. Edinburgh University Press.

12. Dickson, J.H. 2011. *Ancient Ice Mummies*. The History Press; Stroud. Dickson, J.H. 2011. Why Ancient Glacier Mummies are so special: The Ingesta are encoded Maps and Diaries. *Yearbook of Mummy Studies* 1, 45-50. See the author's web pages icemummies.co.uk

13. Dickson, J. H. 2000. Bryology and the Iceman: Chorology, Ecology and Ethnobotany of the Mosses *Neckera complanata* Hedw. and *Neckera crispa* Hedw. Pages 77-88 in Bortenschlager, S. and Oeggl, K. (eds) *The Iceman and His Natural Environment.* The Man in the Ice Volume 4. Springer: Vienna. *Dickson, J.H., Oeggl, K., Holden, T., Handley, L.L., O'Connell, T. and, Preston, T. (*2000) The omnivorous Tyrolean Iceman: colon contents (meat, cereals, pollen, moss and whipworm) and stable isotope analyses. *Philosophical Transactions of the Royal Society of London* 355: 1843-1851

14. Oeggl, K., Kofler, W., Schmidl, A., Dickson, J.H., Egarter Vigl, E. and Gaber, O. 2007. The reconstruction of the last itinerary of Ötzi, the Neolithic Iceman by pollen analyses from sequentially sampled gut contents. *Quaternary Science Reviews* 26, 853-861.

15. Dickson, J.H., Hofbauer, W., Porley, R., Schmidl, A., Kofler ,W. and Oeggl, K. 2009. Six Mosses from the Tyrolean Iceman's Alimentary Tract and their Significance for his Ethnobotany and Events of his Last Days. *Vegetation History and Archaeobotany* 18, 13-22. Dickson, J. *H.* 2013. Bogmoss in the Iceman's Stomach. *The Bryological Times* 137, 28-29.

16. Lorentzen, B. and Rørdam, A.M. 1989. Investigation of the Faeces from a Mummied Eskimo Woman. Pages 139-143 in Hart Hansen, J.P.. and Gulløv, (eds) The Mummies from Qilakitsoq – Eskimos in the 15th century. *Meddelelser om Grønland Man and Society* 12, 1-198. Dickson, J.H. and 12 others. 2004. Kwäday Dän Ts'ìnchí, the first ancient body of a man from a North American glacier: reconstructing his last days by intestinal and biomolecular analyses. *Holocene*, 14: 481-486. Hebda, R., Greer, S. and Mackie, A. 2012. *Teachings from Long Ago Person Found.* Online.

Discovering Gelsolin: And Bringing it from Bench to Bedside

Thomas P. Stossel

Thomas Stossel graduated from Princeton University and Harvard Medical School and trained at the Massachusetts General Hospital and the National Institutes of Health. A practicing hematologist, he has had leadership positions at Harvard-affiliated hospitals, the American Society of Hematology, the American Cancer Society and the American Society of Clinical Investigation. For research on how cells move about the body he was elected to The National Academy of Sciences, The American Academy of Arts and Sciences, and The Institute of Medicine and received honorary degrees from Linköping and Geneva Universities. He is a Founding Scientist of BioAegis Therapeutics and co-founder of a non-profit, Options for Children in Zambia. He is currently Senior Physician at Brigham & Women's Hospital, American Cancer Society Professor, Harvard Medical School and a Visiting Scholar of The American Enterprise Institute. He is author of Pharmaphobia – How the Conflict of Interest Myth Undermines American Medical Innovation *(Rowman & Littlefield, 2015).*

My research work, begun 45 years ago, has concerned a type of white blood cell called a neutrophil. By way of background, every day our bone marrow generates over 100 billion neutrophils that circulate briefly in the blood stream. In a few hours, however, they migrate out of blood vessels into every part of our bodies to find, devour, and kill microorganisms. During this process, each neutrophil moves about an eighth of an inch under its own power: a hundred billion times that distance adds up to more than twice around the earth. Failure to produce neutrophils (as occurs in certain types of blood diseases) or, more rarely, disorders in which neutrophil migration is impaired, result in an inability to control infections with often-fatal results.

My research focused on this migration behavior. Japanese researchers had recently discovered that cells such as neutrophils (and the amoebas they resemble) crawl using the same machinery as found in our muscles. This muscle machinery consisted of fibrous proteins organized in parallel arrays that slide past one another in response to signals from the nervous system to shorten (or contract) the muscle. Exploiting these insights, I decided to try to analyze crawling movements of neutrophils by understanding how these proteins worked inside these cells.

Video: Crawling Neutrophil Chasing a Bacterium - tinyurl.com/crawlingneutrophil

This video is taken from a 16-mm movie made by David Rogers at Vanderbilt University in the 1950s made into a video by Thomas Stossel. It shows a neutrophil surrounded by red blood cells. Bacteria release a chemoattractant that is sensed by the neutrophil. As the polarized neutrophil chases the bacterium through the blood smear, the bacteria is moved by thermal energy in a random path. The neutrophil catches up to and engulfs the bacterium by phagocytosis. The movie is 16x actual speed.

The first goal was to isolate the muscle proteins and see how they functioned. This effort revealed that while in some respects these proteins were similar to their muscle counterparts, they did not operate in parallel arrays. Rather, neutrophil fibers organized themselves into a gel-like substance. This was consistent with observations that dating from the invention of microscopes in the 17th century that the crawling movements of cells seemed to involve transitions between liquid and gel states in their internal substance, like the freezing and melting of water. These transformations enable parts of the crawling cells to be sufficiently rigid and coherent to push against resistance yet at times adequately liquid to allow for changes in shape and direction.

In 1974 my student John Hartwig and I discovered what we named "filamin" that we believed was responsible for the gel formation of the muscle proteins and published the findings in the leading biochemistry and cell biology journals.[1 2] Five years later, another student, Helen Yin, and I discovered a second new neutrophil protein that together with filamin rapidly and reversibly transforms the cellular gels into liquids. We named this second protein "gelsolin."[3]

These discoveries have stood the test of time. The National Library of Medicine repository of research publications as of October 2014 lists 1320 published articles concerning filamin and 2030 about gelsolin.[4] At the same time, a clearinghouse called "ResearchGate" that keeps track of citations to researchers' publications reported that 11,000 such citations referred to mine.[5] I have had continuous grant funding from the NIH since the mid-1970s to do research work related to these proteins. I have won awards, honorary degrees, and election to elite scientific societies in recognition of these efforts. Based on these criteria, I have accomplished what would we could reasonably say is "scientifically important research." But no one has lived one second longer or become healthier due to *direct* results from my "scientifically important" research.

But events *indirectly* related to my research led to a possible invention. In 1981 the unexpected finding emerged that gelsolin is also a protein that circulates abundantly in blood plasma. Cells produce *two* gelsolins – one that resides within them and another that they secrete to the outside world. The latter has come to be called *plasma* gelsolin.

The existence of plasma gelsolin returns us to the fundamental differences between research, invention, and innovation. The identification of plasma gelsolin was a discovery. I intuited that an abundant plasma protein might be medically important and was determined to try to turn this discovery into invention and innovation. But at that point plasma gelsolin's abundance did not guarantee medical importance, because I had no inkling as to its function. Since the academic research culture worships "hypothesis-driven" investigation, and research grant review committees reliably turn down proposals they dub "fishing expeditions" or "descriptive" projects, I had little hope of obtaining any grant support to research the question.

The path around this dilemma required adapting with available resources. At the time I was head of the Hematology and Oncology Unit at Boston's Massachusetts General Hospital and ran a post-doctoral training program for young physicians entering those specialties. Research was a mandatory requirement for specialty certification by the American Board of Internal Medicine, and I had acquired an NIH grant that supported the research trainees. Although many of them had little interest in research, they represented a free labor pool and were relatively happy to participate in a project that might have more clinical potential than the basic cell crawling work that was the mainstream effort in my laboratory.

Since I knew that plasma gelsolin, like cellular gelsolin, binds muscle proteins that are among the body's most abundant, I reasoned that injury might result in plasma gelsolin depletion. Normally, muscle proteins only reside inside of cells, but if the cells' membrane barriers broke, plasma gelsolin could flow into the cells. Plasma gelsolin might also bind muscle fibers released into the blood and help to clear them from the circulation. Measurements of plasma gelsolin in animals and humans subjected to extensive trauma confirmed these theories but did not reveal whether plasma gelsolin depletion might be clinically important. The most plausible theory was that muscle proteins released from injured cells might somehow be toxic and that by removing them, plasma gelsolin had a protective effect.

This theory might explain a common clinical problem. Patients admitted to hospitals for treatment of diverse problems such as trauma, burns, or infections frequently but unpredictably develop a devastating set of complications over the course of hours or days. The affected patient's blood pressure falls, so that blood does not adequately circulate, and vital organs become dangerously starved for oxygen. The patient's lungs fill up with fluid that interferes with oxygen transfer from the air to the blood, further compromising organ viability. These so-called "critical care complications" (that sometimes goes by the name "Adult Respiratory Distress Syndrome") require patients to be cared for in intensive care units where they receive drugs that raise their blood pressure and artificial ventilators force oxygen to their lungs under pressure through tubes inserted into their airways. These measures have side effects, and about a quarter of affected patients die. Others languish for weeks in intensive care. If they survive, they suffer debilitating long-term complications.

If muscle protein released from damaged tissues caused these consequences, and if gelsolin, by mitigating muscle protein toxicity, could prevent these critical care complications, it would be of great medical importance. In the USA, 250,000 patients die annually due to critical care complications, and the estimated cost of critical care consequences is $17 billion.[6]

Research over the next decade provided some, although not overly convincing evidence for the muscle protein toxicity theory.[7] When I pitched to companies the idea that giving plasma gelsolin to patients with low blood concentrations of it might prevent critical care complications, they rightly responded that I had no evidence at all to support that idea. But in 1993, I learned about a protein therapy for the inherited disease, cystic fibrosis. Cystic fibrosis patients suffer progressive lung destruction because their airways fill up with sticky secretions. Scientists know that the stickiness of these obstructions are due to a high content of DNA fibers. The new protein treatment, "Pulmozyme," breaks down these DNA fibers when inhaled by patients, slowing their lungs' progressive deterioration.

Anticipating that pathological cystic fibrosis secretions might contain large quantities of muscle protein from dead neutrophils in addition to DNA and that gelsolin might reduce the thickness of that material, we obtained expectorated lung secretions from cystic fibrosis patients. Experiments revealed that cystic fibrosis sputum contained such proteins and addition of plasma gelsolin reduced the sputum's high stickiness. We reported the discovery in an article in the journal *Science,*[8] filed additional patent applications, and my hospital licensed an invention – to treat airway inflammation with inhaled plasma gelsolin – to Biogen in 1995 for clinical development.

Within three years Biogen produced large quantities of plasma gelsolin using genetically engineered bacteria, documented its identity with gelsolin isolated from blood plasma and showed that giving it to experimental animals in large amounts produced no side effects. Instilled into the airways of normal volunteers, gelsolin also caused no untoward complications. Biogen then performed a small clinical phase II trial administering plasma gelsolin into the airways of patients with cystic fibrosis. The patients who received increasing doses of plasma gelsolin showed a modest improvement in lung functions. But then Biogen decided to drop the project.

The main reason for this decision was that Pulmozyme had recently failed in clinical trials of chronic bronchitis. Chronic bronchitis is an inflammatory disease that predominantly affects heavy smokers and leads to lung destruction. It is far more prevalent than cystic fibrosis. Unless plasma gelsolin was spectacularly superior to Pulmozyme, an unlikely eventuality, competing with an established expensive drug in a small patient population was not economically viable.

Amazingly, the same week that Biogen decided to stop the plasma gelsolin project, a medical journal publication reported that the prognosis of patients suffering from acute trauma could be predicted based on their admission plasma gelsolin levels. All the patients had levels below normal, but the patients with the lowest plasma gelsolin concentrations had a higher probability of critical care complications, consistent with predictions we had made during the previous decade.[9] Over the next several years evidence accumulated that critically depleted plasma gelsolin values precede and predict adverse outcomes in acute and chronic diseases.[10] Additional research revealed that one of plasma gelsolin's functions is to localize inflammation, explaining how it might prevent critical care complications after injuries. Most importantly, giving plasma gelsolin to acutely injured animals reduced their mortality.[11]

In 2005, James Fordyce, a semi-retired investor, became intrigued with the plasma gelsolin project and started a company, Critical Biologics Corporation (CBC), to move it forward. He recruited an experienced biotechnology executive, Ashleigh Palmer, to run the company. Fordyce, Palmer, and I pitched our story to venture capital firms, racing the clock imposed by the patent costs that Brigham & Women's Hospital was now reluctantly footing. In late 2005, an investment firm based on a Hong Kong real estate fortune agreed to finance the company with $10 million in initial funding.

Within two years, CBC reproduced the results obtained previously in the academic laboratories documenting that plasma gelsolin level depletion precedes and predicts critical care complications. It then completed a randomized placebo-controlled clinical trial treating acutely ill patients in an intensive care unit with increasing doses of plasma gelsolin. The purpose of the trial was to determine whether it was feasible to increase sick patients' plasma gelsolin levels without causing side effects. Because of the investor's origin, the trial took place at Queen Mary's Hospital in Hong Kong. The study was a "phase Ib/2a" trial, because the patients were not healthy volunteers as is the rule in phase I trials; the Hong Kong regulators did not consider increasing plasma gelsolin levels above normal in healthy individuals to be ethically warranted.

Although the study was too small to reveal any survival differences between treated patients and those given placebo, it showed that even the lowest plasma gelsolin dose increased the depleted levels in the treated patients. The safety board monitoring the trial concluded that the treatment did not increase adverse outcomes, even though, as is the rule for critical care patients, such outcomes occurred. During the trials, we had also filed additional patent applications for treatment uses including kidney, neurologic, and chronic inflammatory diseases.

In early 2008 CBC set out to raise $24 million to conduct a phase II "proof of concept" clinical trial in the US to document whether plasma gelsolin measurements and replacement of depleted plasma gelsolin would save lives in critical care patients. The cost estimate was based on having to manufacture additional supplies of plasma gelsolin, obtain regulatory approval, and, based on estimates of critical care adverse event rates, enroll, treat, and monitor 400 patients – 200 receiving plasma gelsolin and 200 a placebo.

Our timing was impeccable. As we sought this funding the great recession hit, casting a pall over all biomedical investment. Worse, in terms of funding we were stuck in the middle between the early stages when investors make relatively modest commitments and the later stages when investments are more substantial, but the start up is close to providing actual revenues. By early 2010, CBC's money ran out, and Ashleigh had to find another job. The A-round funder wrote CBC off, the company ceased to exist, and the patent portfolio returned to Brigham & Women's Hospital in June of 2010.

As CBC wound down, Dr. Susan Levinson and Valerie Ceva, two former Novartis Company employees and Steve Cordovano, a retired investment manager had formed a consulting company named Trivalent Partners. They became excited by the gelsolin technology and volunteered to take it forward. The Trivalent team incorporated a new startup, BioAegis Therapeutics. As of early 2014, the BioAegis management had raised over $7 million and made several value enhancing contributions. They convened a panel of critical care physician experts to help design a clinical trial to determine whether detection of critically depleted plasma gelsolin and its replacement will protect patients with pneumonia from complications such as death. The management team hired a contract manufacturer to produce plasma gelsolin and set out to establish a proprietary test to measure plasma gelsolin during the trial and for subsequent clinical applications. They also networked with federal military and bioterror agencies to obtain non-dilutive investment. Since trauma and critical care issues are rampant in war, the military relevance is clear. An infection specialist demonstrated that plasma gelsolin helps white blood cells ingest and kill various types of bacteria, including species that terrorists might turn loose to infect populations.[12] Finally, the team identified a number of other diseases in addition to those previously considered that might benefit from plasma gelsolin diagnosis and treatment. Now the challenge is to see if better economic times and the advances made in the past few years will accommodate raising the funds to run the critical trial described above.

The *research* on gelsolin has gone on for over 33 years. Few researchers have taken projects all the way from "the bench to the bedside." If plasma gelsolin measurement and replacement becomes an innovation, I will have made that complete journey.

My experience illustrates how persistence and adaptation in the face of random events is just as important to the innovation cycle as academic purity or the blessings of research evaluation committees. The projects also illustrate the bidirectional flow between bench and bedside – that innovation is just as likely to promote discovery as *vice versa.*[13]

The companies described in this book that have survived by transforming their business models, merging, taking advantage of other financial strategies, and dropping projects in response to circumstances and exploiting ever changing opportunities, show that innovation is not the result of an idealized linear progression originating from elegant, “innovative” “early-stage” “important” research to product development, product sales, and health benefits. A variety of companies accommodate projects of vastly different natures and scale. In my opinion the only antidote to the failure rate that impedes innovation is to have as many companies as possible taking as many shots on the innovation goal as they can afford. Anything that keeps such companies alive and financially viable contributes to that valuable multiplicity.

Excerpted from "Pharmaphobia' by Thomas P. Stossel

References

1. JH Hartwig and TP Stossel, "Isolation and Properties of Actin, Myosin, and a New Actin-Binding Protein in Rabbit Alveolar Macrophages," *J Biol Chem* 250(1975); T P. Stossel and J H. Hartwig, "Interactions of Actin, Myosin and an Actin-Binding Protein of Rabbit Alveolar Macrophages. Macrophage Myosin Mg++ -Adenosine Triphosphatase Requires a Cofactor for Activation by Actin.," ibid; TP Stossel and JH Hartwig, "Interactions of Actin, Myosin and an Actin-Binding Protein of Rabbit Pulmonary Macrophages. Ii. Role in Cytoplasmic Movement and Phagocytosis," *J Cell Biol* 68(1976).
2. E A. Brotschi, J H. Hartwig, and T P. Stossel, "The Gelation of Actin by Actin-Binding Protein," *J Biol Chem* 253(1978); J H. Hartwig and T P. Stossel, "Cytochalasin B and the Structure of Actin Gels.," *J Mol Biol* 134(1979).
3. H L. Yin and H L. Stossel, "Control of Cytoplasmic Actin Gel-Sol Transformation by Gelsolin, a Calcium- Dependent Regulatory Protein," *Nature* 281(1979); H L. Yin and T P. Stossel, "Purification and Structural Properties of Gelsolin, a Ca^{2+}-Activated Regulatory Protein of Macrophages," *J Biol Chem* 255(1980); H L. Yin, K S. Zaner, and T P. Stossel, "Ca^{2+} Control of Actin Gelation.," ibid.

4. Entrez Pub Med, "Filamin," http://www.ncbi.nlm.nih.gov/pubmed?term=filamin; "Gelsolin," http://www.ncbi.nlm.nih.gov/pubmed?term=gelsolin.

5. https://www.researchgate.net/profile/Thomas_Stossel. Accessed: 9/25/2014.

6. AW Husari et al., "Relationship between Intensive Care Complications and Costs and Initial 24h Events of Trauma Patients with Haemorrhage," *Emerg Med* 26(2009); SV Desai, TJ Law, and DM Needham, "Long-Term Complications of Critical Care," *Crit Care Med* 39(2011).

7. WM Lee and RM Galbraith, "The Extracellular Actin-Scavenger System and Actin Toxicity," *N Engl J Med* 326(1992).

8. CA Vasconcellos et al., "Reduction in Viscosity of Cystic Fibrosis Sputum in Vitro by Gelsolin," *Science* 263(1994).

9. KC Mounzer et al., "Relationship of Admission Gelsolin Levels to Clinical Outcomes in Patients after Major Trauma," *Am J Respir Crit Care Med* 160(1999).

10. MJ DiNubile et al., "Prognostic Implications of Declining Plasma Gelsolin Levels after Allogeneic Stem Cell Transplantation," *Blood* 100(2002); P-S Lee et al., "Relationship of Plasma Gelsolin Levels to Outcomes in Critically Ill Surgical Patients," *Ann Surg* 243(2006); P-S Lee et al., "Plasma Gelsolin, Circulating Actin, and Chronic Hemodialysis Mortality," *JASN* In Press(2009).

11. P-S Lee et al., "Plasma Gelsolin Is a Marker and Therapeutic Agent in Animal Sepsis," *Crit Care Med* 35(2007).

12. Z Yang et al., Plasma gelsolin improves lung host defense against pneumonia by enhancing macrophage NOS3 function, *Am J Physiol Lung Cell Mol Physiol* 309 (2015)

13. I Cockburn and R Henderson, "Public-Private Interaction in Pharmaceutical Research," *Proc Nat Acad Sci USA* 93(1996).

How I Came to Study Ecology

William H. Schlesinger

After many years on the faculty at Duke University, Schlesinger became President of the Cary Institute of Ecosystem Studies, in Millbrook, New York in 2007. He studies global chemical cycles and how they are affected by human activities.

I had no good reason to grow up to be a scientist, especially an ecologist. I was born into an upper-middle-class suburban family on the east side of Cleveland, went to a private school in Shaker Heights, and by all measures was expected to follow my father and grandfather into a career in medicine at Case-Western Reserve University. We never went camping. My mother was deathly afraid of wild things—rodents, poison ivy, and bugs—many of which I brought home with some regularity. I am sure that both of my parents were worried, when with a poor performance in first grade, my teacher commented that I spent a lot of time looking out the window, longing to be outside.

For me, the journey to being a scientist depended on people—people I encountered as role models, mentors, and teachers. Three people were seminal: Russ Hansen who led the weekend Future Scientists Program at the Cleveland Museum of Natural History; a neighbor, Victor Bracher, who worked as a professional marksman for Remington Arms; and Joe Chadbourne, my biology teacher at University School, a prep school in Cleveland's suburbs.

Fresh from a masters program at Michigan State, Hansen was infectious in his enthusiasm for understanding nature, even in the rapidly suburbanizing area of northern Ohio. I saw my first woodcock on an evening camping trip with Russ' class, and I have been a birdwatcher ever since. Even today, I am a big supporter of the role of natural history museums and other non-profit groups in fostering environmental education. I am also a big supporter of field trips; some of my best memories of doing science are associated with sleeping out under the stars.

Video: American Woodcock -- Scolopax minor - *macaulaylibrary.org/video/476611*

Vic Bracher was every ounce a sportsman—an avid game and duck hunter and decoy carver, with a keen eye towards natural history. With my father making long rounds at the University Hospital, I spent many a weekend with Vic around guns and game. He could throw a quarter into the air and shoot a hole through it with a 22-caliber pistol. But, he really enjoyed and I fondly remember sitting in duck blinds with him at dawn, watching and listening to the sounds of a Lake Erie marsh (Video: Sounds of Magee Marsh - http://tinyurl.com/soundsmagee) as it woke up to the new day. I fear his model of the sportsman/naturalist is rapidly disappearing from society, as today's youth spends so much time with computers and not in nature. I also fear that conservation biologists don't appreciate the role of sportsmen in preserving natural habitat for their sport.

As a biology teacher at University School, Joe Chadbourne encouraged any latent enthusiasm and talent he saw amongst his students —be it the mesomorphic students he coached at track and field or science nerds, like me, who were interested in nature. In the summer of 1967, Joe ran a program in ecology for inner-city youth from Cleveland, and hired me as a teaching assistant. On the side, I made a survey of the vegetation of a new campus site for University School, which was published in the *Ohio Journal of Science* with the encouragement of Russ Hansen, and the journal's editor, Jane Forsyth.

News of my interest in entering Dartmouth as a geology major in 1968 was met with a rather cool reception at home. After all, medical doctors were never unemployed, and if drafted to military service, they entered as Captain. Apparently my father was much more worried about Vietnam than I was.

But, damn, Introductory Geology really was interesting. So I took a bunch of geology and geography courses along with various biology classes that could cover the pre-med requirements as well as my interests in ecology. A grounding in both the geological sciences and biology has made it possible for me to become a biogeochemist—a rather new science discipline that focuses on how life controls the chemistry of the surface of the Earth.

In my junior year, a young professor at Dartmouth, Bill Reiners, pulled me into his laboratory for a summer project to measure trace metals in the rainfall of New Hampshire. To analyze lead and cadmium contents, I had to concentrate field samples by evaporation, which gave me a lot of time to read from a pile of old *Science* magazines in John Copenhaver's biochemistry lab. We found lead concentrations were about 50x greater in New Hampshire than in California, which receives clean Pacific air, and linked those to the use of leaded gasoline across the United States. The removal of lead from gasoline has lowered the blood lead levels for urban children across the U.S. to their better health.

Atomic absorption spectrophotometry was fascinating, so I proceeded to analyze all kinds of things for lead—small mammals from the Hubbard Brook Experimental Forest, pencils from Hanover Elementary School, and rocks from Mount Moosilauke. We found huge concentrations of lead in the yellow paint on pencils, alarming a professor in Dartmouth Medical School and eventually leading to the removal of those pencils from the local schools. When it comes to statements on health, an M.D. has a lot more impact on the public than an undergraduate.

Like many college campuses, Dartmouth was abuzz with activism in the early 1970s, and I was elected to chair Dartmouth's first Earth Day program on April 22, 1970, organized by the Dartmouth Outing Club. I like to remind students today that 45 years ago I had a ponytail.

As I worked in his lab, I watched Bill Reiners as he struggled to produce a paper on the carbon content of soils for one of the first scientific conferences on the global carbon cycle, organized by George Woodwell at Brookhaven National Laboratory. A few years later as a PhD student at Cornell, I found that the agronomy library had just what Bill was lacking at Dartmouth—a wide range of journals with reports of the carbon content of soils from the tropics to the polar realm. I produced my first paper on soil carbon in 1977, in the *Annual Review of Ecology and Systematics.* The work showed that the amount of carbon in the world's soils was at least twice that found in Earth's atmosphere and about three times that found in vegetation. Small changes in soil carbon content could have a big impact on the carbon dioxide content of the atmosphere. At the time, I completely underestimated the future impact of the work. Many of us now fear that the large pool of carbon in arctic soils is labile to global warming and conversion to atmospheric CO_2.

I tackled a variety of estimates of the role of soils in the global carbon cycle—how much carbon is lost when soils are disturbed (1986), how much is carried to the sea in rivers (1981), how much carbon is held in desert soil carbonates (1982), and how fast does carbon accumulate in newly deposited soils (1990). Soil carbon is slow to accumulate and fast to decompose when environmental conditions change. Soils rich in humus are not just good for farming and gardening; they store a lot of carbon that might otherwise be in the atmosphere. Being an expert on the pool of carbon in soils got me invited to all sorts of conferences on the global carbon cycle, where I could rub elbows with luminaries who were increasingly concerned about how the human impact on atmospheric CO_2 would affect climate. All now agree that CO_2 emissions from burning fossil fuels, coupled to emissions from human disturbance of forests and soils, are leading to unprecedented levels of CO_2 in our atmosphere, at least for the past 20 million years, and rapid changes in Earth's climate. CO_2 is one of several "greenhouse gases" that absorb infra-red radiation leaving the Earth's surface. Without these gases in the atmosphere, our planet would be very cold, but simple laws of physics show that increasing concentrations of greenhouse gases will produce a warmer planet. Since the distribution of food crops, the height of sea level, and the distribution of disease are determined by climate, changes in global climate are very likely to have large impacts on humans.

Even now, my work at the global scale is unsettling to many of my statistically-oriented colleagues, who want field experiments, replicates, and an analysis of variance. I'm sorry, but you just can't do that with planet Earth—which is one of a kind. Yet, there is clearly a certain amount of carbon that moves from soils to the atmosphere each year, and we can estimate it and compare that number to the amount of carbon dioxide released from fossil fuel combustion. Same too, for global estimates of the flux of ammonia, nitrous oxide or boron to the atmosphere, which I have tackled in more recent years. My work on ammonia focused attention on its emission to the atmosphere during the application of fertilizer and the decomposition of excreted urea in soils. Ammonia is one of several gases, whose concentrations in the atmosphere have changed as a result of human impacts on the global nitrogen cycle.

As a professor at Duke, most of my graduate students focused on field experiments, while I had a fine time occupying myself in extrapolations of local measurements to the global scale. Along the way, I was involved in two large field experiments. The first, the Jornada Basin Long Term Ecological Research (LTER) program sought to ascertain the role of soil heterogeneity in the desertification process. My work showed that "desertified" soils were characterized by a patchy distribution of soil nutrients that exacerbated the persistence of thorny shrub-dominated vegetation and the loss of native grasslands.

The second, the Duke Forest Free Air CO_2 Enrichment (FACE) project exposed large plots of a southern pine forest to the carbon dioxide levels expected mid-21st-century, to see how the trees and soils responded in growth, water use, and soil carbon storage. Our group found that the pine trees grew about 15% faster when exposed to higher CO_2, but there were only small changes in the storage of carbon in soils beneath the forest. Thus, we can expect that forests may partially mitigate the rising levels of CO_2 in the atmosphere, but not enough by themselves to solve the problem of global climate change. Working in large groups has some advantages, but my experiences made me realize that the science that I am most proud of is the work I did and published on my own. I tell that to many students today—don't get lost as a subsidiary coauthor fulfilling someone else's agenda.

My work in deserts and with plant response to high CO_2 both led to several opportunities to provide expert testimony to Congress, other government agencies, and Vice President Al Gore. Being able to offer your science to real problems that face communities is an opportunity not to be passed over by even the most serious laboratory scientist. The challenge is, of course, to explain the work so it can be understood by the general public, and yet to make sure that the gist of the work is still correct. One must focus on what is known, and not dwell on what is unknown—which is what we normally like to do as scientists. And, without being an advocate, scientists should present their work so that we can understand what a prudent course of action would be.

Over the years, I have spent a lot of time "in the field" with some great naturalists—especially Russ Hansen, Peter Marks, Bruce Mahall and Jeff Pippen—and during my own pursuits as a birdwatcher with my wife, Lisa. I was never known as a field botanist, but I thoroughly enjoy memories of field work ranging from the wilds of the Okefenokee Swamp in Georgia to the scorching conditions of the Mojave Desert in California. I fear that as the field of ecology is increasingly dominated by computer jockeys, we will need to remember that there are real species in nature that depend on our studies of their life history traits to inform the best management for their preservation in an increasingly crowded world.

It is easy to say that my proudest accomplishment is found in a textbook, *Biogeochemistry: An analysis of global change*, now in its third edition (Elsevier, 2013) and most recently coauthored with Emily Bernhardt of Duke. This was a chance to put it all together—to see how life has shaped the chemistry of our planet, how the Earth's characteristics have changed through geologic time, and what to expect for the future, as the human population and its resource demand continues to expand. We cannot be cowboys in the modern world; there are simply too many of us. And while it might be nice to have the freedom of cowboys to do what we want, the planet can no longer sustain that behavior if we are all to survive at minimal standards of living and at peace.

Discovering the Beauty of Sea Slugs

Ángel Valdés

Ángel specializes in the systematics and evolution of sea slugs. He obtained a Ph.D. in Biology (Zoology) from the University of Oviedo in Spain, and conducted his postdoctoral training at the California Academy of Sciences in San Francisco. He is currently an Associate Professor of Marine Evolutionary Biology at the California State Polytechnic University in Pomona. During his career he has published over 100 peer-reviewed papers describing hundreds of new species to science and published several books on the diversity of sea slugs.

Sea slugs are among the most beautiful organisms in our planet. The bright, contrasting color patterns of many sea slug species rival those of butterflies, tropical birds, and dart frogs, among others. As in all these other groups of animals, the conspicuous colors of sea slugs are a warning, by which the sea slugs signal to potential enemies that they are either poisonous or venomous. Many sea slugs feed on algae or on other animals that are loaded with toxic chemicals. Sea slugs are immune to the toxins and have evolved the ability to accumulate those chemicals in their own bodies making themselves poisonous to other animals. Other sea slugs have the ability to synthesize their own defensive compounds completely independently from their prey or by chemically transforming ingested compounds into more potent forms.

But the most remarkable example of defensive mechanism in sea slugs is present in species that feed on cnidarians (animals like sea anemones, corals, sea jellies, etc). Cnidarians contain stinging cells that are used to inject venom into potential predators or prey. Some sea slugs can steal the stinging cells from their cnidarian prey and utilize them for their own defense, making themselves venomous. This is called cleptodefense, or stolen defense. Although most sea slugs sequester stinging cells too small to penetrate the human skin, some species that feed on particularly nasty cnidarians can inflict painful stings to people.

Some of the toxins found in sea slugs are useful in medical research and a few are currently being tested in clinical trials against certain types of cancer. But sea slugs are important to humans in several other regards. Some sea slugs are model organisms in neuroscience and have been critical in the discovery of how memories are formed and learning takes place in simple brains. (*see* A Nobel Prize with help from sea slugs - http://tinyurl.com/kandelmemory and A Quest to Understand How Memory Works - http://tinyurl.com/questmemory). Other species have been proposed to control the spread of invasive seaweeds, or have became invasive themselves, causing environmental problems. One of those invasive species, *Haminoea japonica*, is known to carry a schistome parasite that can penetrate the human skin causing painful rashes to swimmers in the San Francisco Bay (http://journals.plos.org/plosone/article?id=10.1371/journal.pone.0077457).

Some sea slugs have been traditionally used as food by native people in a few places in the world, but most species are too toxic to be edible. In fact, the Romans were known to use extracts from sea slugs to kill their political enemies. It is important not to confuse sea slugs with sea cucumbers, which are used in Chinese cuisine. Sea slugs are mollusks, related to sea snails and terrestrial slugs and snails, whereas sea cucumbers (sometimes incorrectly called sea slugs) are echinoderms, related to sea urchins and sea stars. Another common name for sea slugs is nudibranch, but actually nudibranchs are a group of sea slugs and not all sea slugs are nudibranchs.

Possibly the most amazing aspect of the biology of sea slugs is the ability of some species to photosynthesize. It is well known that some species of marine animals, such as corals and sea anemones contain microscopic algae in their tissues, called zooxanthellae. The animals benefit from the sugars produced by the algae and in turn the algae obtain protection and a stable environment from the animals. This is called a mutualistic relationship. A few species of sea slugs can sequester the zooxanthellae from their prey and can keep them alive in their own tissues benefiting from their photosynthetic activity. But another group of sea slugs, called sacoglossans, go well beyond that. Sacoglossans feed directly on seaweeds by piercing the tissue using blade-like teeth and sucking up the internal contents. Among those contents are chloroplasts, the organelles that allow plant cells to photosynthesize. Many sacoglossans can incorporate those chloroplasts into their own cells and keep them alive for variable periods of time. We all learned in high school biology that only plant cells can carry out photosynthesis, well, this is wrong! Both plant and sea slug cells can photosynthesize! (*see* http://jxb.oxfordjournals.org/content/64/13/3999.long)

I first became interested in sea slugs at a young age. I grew up in a small village by the Cantabrian Sea, in Northern Spain, and I used to spend my weekends looking for all kinds of animals and plants in the tidepools. I remember very well my first sea slug, a beautiful navy blue and yellow species I discovered underneath a rock. That single event changed my life in more profound ways that I ever imagined. During the rest of my life I engaged in a quest to discover as much as I could about these fascinating organisms, their diversity and evolution.

One of the most surprising aspects of the biology of sea slugs to me was how little we knew about them. Although nearly 6,000 species have been documented and described to date, probably the same number of species remains to be formally named. How was it possible that so many incredibly beautiful animals have not yet been documented by science? I quickly learned that the process of naming species required a considerable amount of training, hard work, and sometimes luck. I decided I wanted to make a contribution to this process of discovery and documentation of the diversity of life. My quest for knowledge took me to several places around the world, after getting my PhD in the Universidad de Oviedo in Spain, I moved to Paris, France to work on the sea slug collections at the Museum National d'Histore Naturelle. There I worked with some specimens collected during the 1700s and 1800s and studied by legendary naturalists such as Cuvier and Lamarck. Subsequently I came to the US to work at the California Academy of Sciences in San Francisco where I learned modern methodologies to study sea slugs, such as molecular biology and scanning electron microscopy. And finally at the age of 30, I obtained my first real job, as the mollusk curator at the Natural History Museum of Los Angeles County, where I was finally able to establish my own laboratory and became an independent researcher.

During this time, I was also able to conduct fieldwork, searching for sea slugs around the word, including countries well known for their biological diversity like Costa Rica and the Philippines, but also more exotic and removed places such as New Caledonia or Vanuatu.

Discovering new species of sea slugs is an extremely rewarding process. It begins with the excitement of the initial finding in the field, encountering an animal that no one has ever seen before. The specimens are then preserved and taken to the laboratory for further study. The specimens are then dissected under a regular microscope and certain parts, such as the teeth are examined under a powerful scanning electron microscope. These data may provide definitive evidence that this animal is distinct from closely related species. However, sometimes the evidence is inconclusive and further research is necessary. In that case molecular work is very important. We can read the DNA of the specimens and compare it to that of closely related species; this often provides conclusive evidence that the new species is actually different, but also allows us to put the new species in an evolutionary context of classification. Using powerful software we can infer the evolutionary relationships of the new species and even determine the time in which the new species split from close relatives. The final step is publication, when one can give a new name to the new species and your own name becomes immortalized in conjunction with the species name.

I do not keep count of how many species of sea slugs I have described during my career, but the number is probably around 200; this may seem impressive to some people, but it is actually a tiny percentage of the total diversity of sea slugs. I am particularly fond of several sea slugs I named. For example, I named a new species from Florida after my wife Stephanie, called *Berghia stephanieae*, and in collaboration with my postdoctoral supervisor, Dr. Terry Gosliner, I named a new genus (and species) after Nelson Mandela (*Madelia mirocornata*). Some of these new species are incredibly beautiful tropical shallow water sea slugs, but I also had the chance to describe many new deep sea and polar species, often not so brightly colored. Among those, are the first sea slugs ever found in strange marine ecosystems such as hydrothermal vents or whale falls. These are chemosynthetic ecosystems, where microorganisms use chemical energy from compounds dissolved in water that seeps from the ocean crust or from the bones of dead whales to grow, sustaining entire ecosystems isolated from sunlight. Specifically, these microbes oxidize molecules of hydrogen sulfide or ammonia to obtain energy, constituting the base of the food pyramid. Other organisms either feed on the microbes or maintain mutualistic relationships with them. The sea slugs found in these ecosystems are carnivores situated very high in the food pyramid. (*see* http://www.eu-hermione.net/science/chemosynthetic-ecosystems)

I also had the opportunity to describe some neustonic species of sea slugs, which live in the intersection between the ocean and the atmosphere. These amazing animals stay afloat by swallowing air, making their bodies buoyant, and protect themselves against damaging ultraviolet solar radiation by having a beautiful blue coloration. (*see photo:* http://eol.org/pages/451180/overview)

Most of this work has been done in collaboration with colleagues in different universities and research centers. In a globalized world, exchange of ideas and data across borders and laboratories is easier and more important than ever. Far from the lone naturalists of the 1800s, such as Darwin and Wallace, who carried out their work and developed their ideas in relative isolation, modern scientists cannot remain competitive in their fields without sharing information and working with colleagues across the globe.

There is only one thing in my work more rewarding than discovering new species of sea slugs, it is to teach other people how to describe new species and share with them the beauty of these organisms. Scientific discovery, as well as other aspects of human creativity, are pretty much useless unless we can pass the new knowledge and ideas generated on to the next generations. This is what makes my current job so incredibly fulfilling. As a professor in the Biology Department of the California State Polytechnic University, I have the opportunity to work with a bright group of students and share with them the excitement of discovery. My students are full of energy and passion for their research projects. My highest achievement is to help them grow as scientists and become more self-confident as they graduate and move on with their careers. It makes me incredibly proud to see them publish their first peer-reviewed paper, or getting accepted in a competitive Ph.D. program, to become part of the next generation of scientists. However, this is not completely an altruistic process, as I often learn more from my students than they learn from me. Their fresh approaches to solve problems in their research and their ability to think out of the box are an invaluable learning experience and an inspiration.

The discovery of new species and the documentation of the diversity of life on Earth are now more important than ever. Our planet is confronting unprecedented challenges. Climate change, habitat destruction, depletion of natural resources, and other human induced processes are causing species extinctions and profound changes in natural ecosystems all over the word. We must understand those species and ecosystems before it is too late to do anything to prevent the irreversible loss of biological diversity. Many of these beautiful species of sea slugs may contain the key to cure human diseases and improve the quality of life of millions of people. We cannot afford to ignore them and simply let them go.

I will never cease to be amazed by the incredible beauty and diversity of sea slugs. But more importantly, I will never get tired of sharing with other people the excitement of discovery and the importance of understanding the natural world.

Into the Unknown

Nils-Axel Mörner

Nils-Axel Mörner is a geologist and geophysicist. He was head of Paleogeophysics & Geodynamics *at Stockholm University (1991-2005). He is a front specialist on sea level changes. Besides hundreds of peer-reviewed papers and some books, he was president of the INQUA* Commission on Sea Level Changes and Coastal Evolution *(1999-2003), leader of the International Sea Level Project in the Maldives (2000-2005) and coordinator of the INTAS project on Geomagnetism & Climate (1997-2003). In 2008, he was awarded* The Golden Condrite of Merit *for his work on sea level changes. As president of the INQUA* Neotectonics Commission *(1981-1989) and editor of the* Neotectonics Bulletin *(1978-1996), Mörner laid the ground for modern neotectonics and paleoseismics. In Sweden, he recorded and dated 61 large earthquakes after the Ice Age, 17 of which were linked to tsunami events. Mörner has also studied paleomagnetism (where he coined "the Gothenburg flip"), Earth's rate of rotation (and its interaction with ocean circulation and regional climate and sea level) and the Solar Wind and its effects on Planet Earth. Mörner has written 605 scientific papers (most of which in peer-reviewed journals), given more than 542 papers at international meetings, and published 12 books. Besides, he is* Patronus Skytteanus *for a foundation in the entailed estate instituted in 1622 by Johan Skytte for a professorship in "*Eloquentia et Politices*" at Uppsala University with a mandate to manage the foundation and appoint the new Professor Skytteanus. From 1994, the foundation awards* The Johan Skytte Prize in Political Science[2].

Discovery is the signum of humans and their evolution through time. Step by step have we widened our field of knowledge; from the highest mountain peak to the deepest abyssal hole, further and further out into macro-cosmos and deeper an deeper down into micro-cosmos, into the laws of physics and biology, from the first traces of art and writing to masterpieces in painting, music and literature. Sometimes progress is hampered by too complicated systems of interacting variables (or simply lack of imagination). Usually, however, we have to add "for the moment", because, with time, most obstacles will be dealt with. Step by step we concur new sectors and widen our knowledge. Still, there remains enormous sectors, where our knowledge and insight are still insignificant or even lacking.

The process of discovery is personal. Its beginning and end differ, and so does, of course, the very discovery itself. Therefore, the personal history may attract some general interest. Below, I will describe my personal way through life, and I chose to highlight one factor – sea level changes – which has remained a central theme for me[3].

Starting the journey through life begins with "packing the backpack," upon which you should feed and survive – that is the childhood, the environment in the family and the education.

Dreams and poetry were woven together into a new reality and happiness. In this remarkable environment my life commenced. Of course, I always came to search, dream and formulate new views and interpretations.

Direction Set

My father (Stellan) was an artist and pioneer for surrealism. He was the oldest son of a noble family in Sweden. He left the castle he was to inherit and built his own castle of happiness and dreams. My mother (Moussia) was "a pre-revolutionary girl" from St. Petersburg, who had to escape with her family to Paris when the Bolsheviks took over in Russia. They both went to the same art-school in Paris, and there they met for lifetime.

No doubts, I had a most lovely childhood. I was brought up among artists and authors. Our summers were spent on the Swedish west coast, where sea and sky met and where surrealistic happenings may always occur. In the backpack I stuffed: unrestrained life and artistic sensibility.

Mother had a remarkable feeling for honesty and standing up for what is true. In the backpack I stuffed: never lie and always stand up for what you believe.

Only a surrealist can depict the expectations for a new year and visualize what still exists only in the hope and dream. In creative science we too extend ourselves into the unknown attracted by visions of new facts and new relationships.

Early education

I went to a boarding school. To begin with, I wasn't a very successful student (partly because of quite serious dyslectic problems). Then I understood that freedom was not to be found in misbehaving and not doing what was expected. So, I started to work hard in school (not easy for a lonely dyslectic) and soon had the best rating in the entire school. From that moment I was a totally free person; and I was able to do whatever I liked. In the backpack I stuffed the notion: if you are the best, you can do what you want.

Sport was a strong part of the boarding school life. I loved it. As a part of all the running, skiing and skating, I became in love with nature itself and all what it offered. In the backpack I stuffed: love of nature and environment.

I was the editor of the school newspaper and the prefect for one of the student houses. In the first case, I learned how to write. In the second case, I learned how to talk and how to help students with poor grades. In the backpack I stuffed: how to write and especially a love of speaking and convincing.

Also, there were many good friends. One of them, Bjarne Lembke (later medical doctor and agronomist) was indeed another "freethinker" and we became friends for life. He took part in my marine geological sampling of the seabed in the Kattegatt Sea (for my thesis in the 60s), in the sea level project in the Maldives (2000, 2005) and in the work in Goa, India (2011).

So, I left primary school with a quite well stuffed backpack for the life to come. Besides the above-mentioned content in the backpack, I had a firm self-confidence to rely on. Also, I had learned how hard work paid-off. One problem though; I was very “blue-eyed” and totally inexperienced in the evilness awaiting in the adult world. Another problem was that sport had taught me fair play – and this was certainly not what was waiting in the future.

In 1961, I married a remarkable girl. How I won her, I cannot understand: the most smashing beauty along the west coast of Sweden, with humor and deep intelligence. Without Ulla at my side, I don’t know where I had ended up. Children followed 1963 (Jonas), 1966 (Ninna) and 1976 (Philip). The family harmony gave me an unbeatable backbone in coming fights.

University calls

Of course, I should go to university. There was not even an option for me. But what subject?

To begin with, I was considering religion, but, luckily, switched to zoology. A very hectic time started were I worked like hell; studied several different topics at the same time, wrote bachelor theses both in physiology (synchronization of cells by cold shocks) and geology (studies of a Devonian Crossopterygian fish fauna from northern Canada) and taught biology in the evenings. The work in physiology, though very interesting and useful, made me feel too much like a laboratory rat. Even the fish fauna was too dusty and away from real life. I came to stumble over Quaternary Geology[4] and became trapped there for lifetime.

Quaternary Geology provides an attractive combination of hard work for the body in the field and deep intellectual work in interpretation and conclusions. It fitted me perfectly well. Through out my life, it has become a “modus operandi” for me: very hard work in the field (often directly ascetic with respect to food and comfort) and deep intellectual handling of the interpretative and concluding parts. From Quaternary geology, I came to drift over into geophysics (paleomagnetism and paleoseismology) and partly even physics (Solar Wind interaction with the Earth and planetary influence on the Sun).

As a reasonably young scientist I got a personal position as an associate professor at the Swedish Research Council in Paleogeophysics and Geodynamics[5]. This position was later transformed into an independent research unit at Stockholm University which I headed from 1991 to my retirement in 2005. It grew to a strong and well-known international institute. We were in the front when it concerned sea level changes, paleoseismology and paleomagnetism. I conducted research practically all over the globe. Ten students took their Ph.D.-examinations at the unite of Paleogeophysics & Geodynamics (P&G). Numerous people came from all over the world to work with us. The front geophysicist in Russia came for a 1-year project but stayed for 4 years. Several major international meetings and field trips were conducted. Numerous papers were published, and communications given at major international meetings.

Sea Level Changes

My Ph.D.-thesis (a heavy book of 487 pages and 10 big plates in color) was devoted to the Swedish West coast and the Kattegatt Sea; the mode of deglaciation and the changes in sea level. Because it covers the margin of glacial isostatic uplift[6] after the last Ice Age, it was possible to pinpoint the sea level changes with a very high precision and apply direct dating by the C14-method. With this material at hand – i.e. this detailed spectrum of sea level records over the last 13,000 years – it was possible to separate the isostatic uplift component and the eustatic[7] sea level component. Because of the unusually high number of C14-dates, it became the most detailed and well-dated sea level curve in the world when published in 1969.

In the 1960s there was a very vivid debate on the character of the postglacial sea level rise; especially between Professor Fairbridge advocating an oscillations trend reaching well above the present in Mid-Holocene time[9] (about 4000-6000 years ago), and Professor Shepard advocating a smooth and continual rise up to the present. My curve happened to be something just in between; an oscillating rise of low amplitudes reaching the present in Late Holocene time. As a young scientist I ended up in the strange situation where Fairbridge cited my curve in evidence against Shepard's curve, and Shepard cited my curve in evidence against Fairbridge's curve – and suddenly I was in the front of international sea level research.

At that time, sea level changes were assumed to be "globally synchronous and similar," and this was also the definition of "eustasy." Consequently, with the eustatic factor defined in the Kattegat (NW Europe), all other coasts of the world could be assessed with respect to crustal stability and sea level changes. Coastal uplift or subsidence would make the local sea level changes diverge more and more back in time from the global eustatic sea level rise.

After having analyzed sea level data from sites all around the globe, I came to the surprising conclusion that theory and reality didn't match. Instead of diverging back in time, almost all sea level curves crossed each other at around 7000-8000 C14-years BP. I wrote a paper in 1971 saying that the real Holocene sea level problem was not whether the sea rose like Fairbridge's curve or like Shepard's, but that all curves were more or less together around 8000 BP, which indicated that sea level during the intermediate time must have been differentially distributed (i.e. not parallel) over the globe due to a hitherto unknown sea level factor.

With the satellite era and the beginning of space geodesy, it became obvious that the Earth's hydrosphere was not a smooth ball flattened to the poles (as understood by Newton), but a wrinkled surface in response to internal irregularities in gravity; i.e. the ocean surface had a relief. This gravitational potential surface is termed the geoid. The largest difference is between the high of +86 m over New Guinea and the low of -104 m over the Maldives; in total a difference of 180 m with respect to the center of the Earth. When I saw the first geoid map in 1971, I immediately realized that this was the answer to the "real sea level problem" identified the same year. A new paradigm was born: sea level is irregular and from this follows that changes in sea level cannot be parallel (as assumed before) but have to be irregular, as illustrated below.

This was a totally novel concept that revolutionized sea level research. From now on each region had to define its own eustatic changes. The old concept of global eustasy had to be redefined.

In order to be sure that I hadn't missed anything, I studied a large number of papers and books in geophysics and physics for several months, wrote up a paper and sent it to sharp and clever colleagues. After that process, I became convinced that I was right, and submitted a paper to an international journal of high standing. The reviewing process was interesting. One of them said it was nothing new. The other said that it was a remarkable paper and "if any paper in Earth science would deserve a Nobel Prize, this is it." The paper was published and has become a standard reference in sea level research[10]. And the favorable reviewer, Professor Walter Newman (New York), became a close friend and would come to spend the rest of his life on the question of sea level changes and geoid deformation.

In view of the new concept of global sea level changes the term "eustasy" had to be redefined simply as "changes in sea level" (1986) in opposition to changes in land level[11]. Consequently, each region had to define its own "eustatic curve" (so I did for Northwest Europe in 1980).

The concept of changes in the geoid topography seemed to suggest that sea level data from different parts of the world would differ significantly. A closer study revealed another fact, however. There occurred both positive and negative correlations, which had to be explained in new terms. The Late Holocene sea level changes in East Africa were directly opposed to those in Peru, indicating that water masses were flushing back and forth in east-west direction. Records in South Carolina and in Connecticut fitted the Northwest European sea level data very well, indicating a mutual effect from variations in the transport of water masses along the Gulf Stream. Similarly, there were some correlations between some records from Japan with those in Northwest Europe, suggesting simultaneous pulses in the Kuroshio Current and Gulf Stream.

The driving forces had to be the interchange of angular momentum between the solid Earth and the hydrosphere[12]. The idea was first presented in 1984; 16 pulses in intensity of the Gulf Stream were recorded during the last 10,000 years. A number of subsequent papers followed.

The beat of the Gulf Stream is especially interesting as it brings warm equatorial water masses from low latitudes to high latitudes.

Consequently, it had to affect temperature (measured in paleoclimate), water masses (measured in regional sea level) and the global distribution of mass (balanced by changes in rotation in a feedback coupling).

The question was if those changes were driven by internal oscillations or if they were also affected by external impulses. In Europe, we have a remarkable reconstruction of the changes in temperature, at least, a millennium back in time. In these data, it was possible to record the above-mentioned beat in Gulf Stream activity[13]. During the periods of Solar Minima[13] known as the Dalton, Maunder and Spörer Minima, cold Arctic water penetrated all the way down to middle Portugal and caused Little Ice Age climatic conditions in Europe and the Arctic.

The dislocation of water masses in the northeastern parts of the Atlantic is a function of a change in the rate of rotation. Because it occurred during all the three Solar Minima during the last 600 years, the changes in solar activity was likely to be involved, in one way or another. A new theory was born (1996): changes in Solar Wind affect Earth's rotation[14].

The Solar Wind (ejection of plasma from the Sun) varies in mass and speed with the solar cycles. It interacts with the magnetosphere of the Earth increasing the geomagnetic field and its shielding capacity at solar cycle maxima, and decreasing the geomagnetic field and its shielding capacity at solar cycle minima. This alters the penetration of cosmic rays and ^{14}C production in the atmosphere. Obviously, it also affects Earth's rate of rotation. Finally, we get an explanation of the correlations observed by many scientists between changes in solar activity and rotation (length of day; LOD). Ultimately, the solar variability seems to be driven by a planetary beat[16]. Planetary beat is the gravitational and rotational effects the planets have on the Sun, its motions and activity with respect to emission of luminosity (irradiance) and solar wind.

This reviews my "scientific journey" from detailed sea level studies in SW Sweden, via changes in the geoid topography and changes in the oceanic circulation, to sophisticated analyses of the Earth's rate of rotation and interaction of changes in the Solar Wind. It illustrates how a scientists work, digging deeper and deeper into a subject, formulating new concepts just to open doors to new findings and additional concepts. One, so to speak, closes bolt after bolt to finally get a machine that works.

This type of "journey" is the reward for a scientist. It cannot be accomplished with intellectual calculations, but goes via astonishing "aha-experiences."

Of course, I became "a sea level specialist." Christopher Booker in The Sunday Times (March 23, 2009) kindly wrote: *"if there is one scientist who knows more about sea levels than anyone else in the world it is the Swedish geologist and physicist Nils-Axel Mörner."* In 2008, I was awarded "the Golden Condrite of Merit" from the University of Algarve in Portugal *"for his irreverence and contribution to our understanding of sea-level change."*

With this in hands it was time to challenge the proposition by the Intergovernmental Panel on Climate Change (IPCC) that, due to the atmospheric rise in CO2 and global warming, sea level will rise, flooding low-lying coasts around the world. This is certainly not the case. Sea level has remained virtually stable over the last 40-50 years. By the year 2100, the proposed sea level is likely not to be more than +5 cm ±15 cm.

The separation of the isostatic and eustatic components in the sea level spectrum from the Kattegatt region (1969) became the key to a large number of other factors and questions. The stepwise improvement of the concept of eustasy has been reviewed above. The course and mode of postglacial isostatic uplift of the Fennoscandian shield, the exposed northwest segment of the East European Craton was disclosed; 800 m of uplift in the center, rates ten times higher than plate motions and transfer of mass in a channel flow of low viscosity in the upper mantle of the Earth (which invalidates global loading effects[17]). The rates of uplift initiated a large number of high-magnitude earthquakes. At the time of deglaciation, Sweden was a high-seismic area (and scientists from all over the world came to our field excursions to learn about it).

The implication of my sea level odyssey through time and space is that I am able firmly to state the all the talk about an ongoing disastrous sea level rise[18] is nothing but exaggeration and disinformation[19]. It steals the limelight from truly urgent problems in the world.

Nils-Axel Mörner

Creativity in Operation

What is creativity? I leave that question for the moment and go to another one, which I can answer better; *How does creativity operate?*

In mathematics, we may solve a problem by intellectual efforts; 2 + 2 = 4, and nothing else. In almost all other scientific (and maybe even daily) problems, we feed ourselves with information but we cannot solve the question. This has to be left to the subconscious. There – beyond our intellectual range – the data are processed. And suddenly, out jumps the answer: eureka!, aha!, that's it – suddenly we have the answer. Often it strikes us when we least would expect it; in the car at a red light, during the evening walk with the dog, or at any other trivial moment of your life. This is how creativity operates.

As a matter of fact, there are so many other things in ordinary life, which may be solved in the same way. What should I serve all these people coming for dinner on Saturday? – and suddenly you have it. Not by intellectual deduction but by an "aha" from the subconscious.

My car was my creativity machine. Sitting alone driving, over and over again the struck of lightening was solving a problem. Nowadays, it frequently occurs when I am out with the dog; woauw, that's it!

But one needs to allow oneself to listen to the voice from the subconscious. In some of my courses I try to teach people how to learn to listen to their subconscious. Those wonderful moments of aha-experiences are the reward in science. This is how science goes forward. Those who talk and talk about "intellectual deduction" simply do not know what true creativity is – it is the divine gift from the subconscious.

Now we are starting to hit what real creativity is. Creativity doesn't age. Picasso remained a front person in art all through is life; always changing, never aging. So, did my Father, and so I think I am doing, too. The flashes of aha continue to strike me over and over again (with 113 papers and 8 books after retirement 10 years ago).

It is true that the young generation is the carrier of new techniques; but creativity – certainly not. Creativity is individual and person-specific, and here there are no age limits.

It is sad that research councils and founding agencies seem to have little or no insight to this fact, which we may even regard as a "natural law" or "psychological law."

Bubbling of ideas

Creativity is a constant "bubbling of ideas" from one's subconscious mind. It is like an internal fountain bubbling and splashing when you least expect it. Here the scientist, like the artist, almost becomes a tool of creativity. It is a rather self-going process not seeking merit rating (or sale of a painting) but the solution to a problem (or the visualization of a vision). One paper (or painting) finished, and a couple of others are borne in your mind.

In society, you may become a complicated person who goes your own way. Claiming something new, not claimed before, implies that you may become called "controversial". In art, this is positive, but in science it is negative. This is very strange because it refers to the attitude to novelties. Why should that be negative (or critical) in science? Why should science be more rigid and conservative than art (at least today)? The answer must lie on the personal plane; with new ideas, insights and paradigms, the persons who held the previous (old) views become passed and outdated. Instead of changing opinion and assimilating the novelties, they dismiss them. This is to become old in position. The universities are full of such persons. Sometimes it even seems "self-gravitational"; one mediocrity seeking other mediocrities.

When it goes to founding, one may even say: *"those who have money, have no ideas, and those who have ideas, have no money."* At least, this was how I saw the situation in Sweden. But it doesn't matter. The talented cannot be stopped. He or she continues to work without regard to the difficulties: *talentum omnia vincit* – talent conquers all.

A few papers of mine

If I were to choose just a few papers of mine to recommend, it would be these:

1971 Late Weichselian paleomagnetic reversal – *Nature*, 234, 173-174.

The paper was written on the airplane in early December, submitted to Nature and appearing in print on December 26. A geomagnetic reversal as young as 12,400 BP was sensationally. It became known as "The Gothenburg Excursion" (or "Flip"). The paper has become widely cited through out the years.

1976 Eustasy and geoid changes – *Journal of Geology*, 84, 123-151.
This was a true benchmark paper that completely changed the way of understanding sea level changes. It meant a paradigm shift. It has become a standard reference in sea level research. It led to the redefinition of the term "eustasy" (1986).

1995 Earth rotation, ocean circulation and paleoclimate – *GeoJournal*, 37, 419-430.
This was an elegant summary of my theory of a feedback coupling between changes in rotation and changed ocean surface current circulation (the first paper of which was published in 1984). In a subsequent paper in *An. Brazilian Acad. Sc.* (1996, 68 Supl. 1, 77-94), the influence of Solar Wind intensity changes was for the first time realized.

2003 *Paleoseismicity of Sweden. A novel paradigm.* – A contribution to INQUA from its
Sub-commission on Paleoseismology at the Reno, Nevada, 16th INQUA Congress.
This monograph of 320 pages documented and described 54 paleoseismic events in Sweden after the Last Ice Age. Faults, bedrock fractures, caves, liquefaction event in multiple phases, tsunami events, turbidites spread over extensive areas in one and the same varve-year were recorded. The book shook the scientific ground like a true earthquake (not least the nuclear power industry). No doubt, at the time of deglaciation when land rose at rates ten times faster than plate motions, Sweden was a high-seismic area both in frequency and magnitudes. Today we have 62 events recorded (e.g. *Quaternary International*, 242, 65-75, 2011; *Pattern Recogn. Physics*, 1, 75-89, 2013), where the PRP paper provides a first general account of how to handle seismic hazard assessments over exceptionally long time periods. It invalidates the application of the KBS-3 method proposed as repositories both in Sweden and Finland.

2008 Paleoearthquake deformations recorded by magnetic variables (by Mörner & Sun) –
Earth and Planetary Science Letters, 267, 495-502.
At earthquakes, the magnetic grains get a second chance to rotate and re-align with respect to the geomagnetic field. Consequently, the grains carrying the magnetic remanence must be very small and rather of nano-

size. All this was completely new in paleomagnetism. It provided a new method of discriminating between plastic and liquefied deformations. It may also create crypto-deformations not visibly observed.

2012 Planetary Beat, Solar Wind and Terrestrial Climate – In: *"Solar Wind: Emission,*
Technologies and Impacts", 47-66, Nova Publ. Co. 2012.
Solar Wind, Earth's Rotation and Changes in Terrestrial Climate – *Physical Review &*
Research International, 2013, 3 (2), 117-136.
The new concept was that the Solar Wind (not the solar irradiance) plays a central role in the effects of solar variability on Earth's climate. The ultimate forcing function may come from a gravitational "Planetary beat" (something I proposed 30 years ago, and nowadays starts to be discussed by several scientists). The concept gave rise to the special issue on planetary, solar and terrestrial interaction of *Pattern Recognition in Physics* in 2013[15-16].
The basic planetary-drive theory led to a new book on *"Planetary-Solar-Terrestrial Interaction and a Modern Book-burning"* in the stage of being printed (Nova, 2015).

2013 Sea Level Changes: past records and future expectations – *Energy & Environment*, 24 (3-4), 509-536.
This paper gives a comprehensive review of sea level changes and the frames to be considered in our estimations of possible changes in the near future[19].

Foot-notes

1. This essay will follow one scientist's journey through life and his step-by-step widening of the concept of sea level changes from field studies in Sweden out over the world and even into the effects of planetary beat. In this paper it is not the facts that are in the center, but the ways they evolve, interact and produce new insights and aha-experiences. With the below foot-notes, I provide additional information on the scientific material that might be useful for an insight also to this part of the essay.
2. See: N.-A. Mörner & L. Bennich Björkman, *Johan Skytte, his donation in 1622 and the Johan Skytte Prize in Political Science*, The Skytte Foundation, 1, 1-90 (2014).
   Also: http:/skytteprize.statsvet.uu.se/PrizeWinners.aspx
3. Sea level changes are the function of numerous interacting forces, which must all be considered in order to make meaningful estimates of coastal hazards. Millions of people are living close to the sea. A rapid sea level rise

would be disastrous. But such worries are not based on actual observational facts.

4. Quaternary geology refers to the geology of the Ice Ages during the last 2.5 million years.
5. Paleogeophysics refers to the geophysics of the past, and geodynamics to all kind of processes operation in nature. This implied that we could do whatever we wanted.
6. Glacial isostatic uplift refers to the rise of the Earth's crust as a function of the recovery from ice load during the Last Ice Age.
7. Eustasy refers to the sea level rise as a function of the melting of the ice masses after the Last Ice Age.
8. The ancient shorelines in Fennoscandia are tilted from the center to periphery of uplift.
9. Holocene refers to the last 10,000 C14-years (i.e. the last 11,000 absolute years).
10. See: Journal of Geology, 84, 123-151 (1976).
11. See: Journal of Coastal Research, Spec.Is-1, 49-51 (1986).
12. The hydrosphere id the open water masses of the world; i.e. predominantly the ocean water masses.
13. See: Global and Planetary Change, 72, 282-293 (2010).
14. See: An. Acad, bras. Ci, 68 (Supl. 1), 77-94 (1996).
15. From: Mörner and 18 others, General conclusions regarding the planetary-solar-terrestrial interaction, Pattern. Recogn. Phys., 1, 205-206 (2013).
16. Planetary beat is the gravitational and rotational effects the planets have on the Sun, its motions and activity with respect to emission of luminosity (irradiance) and solar wind. See: *Pattern in solar variability, their planetary origin and terrestrial impact*, Special Issue edited by N.-A. Mörner, R. Tattersall & J.-E. Solheim, Pattern Recognition in Physics, 2013.
17. The global loading model suggest that weight of the ice caps and the de-loading when those ice caps melted after the Ice Age goes through the Earth and affects the entire globe, which neither the mode of Fennoscandian uplift or sea level records in the Pacific and Indian Ocean validate.
18. The idea of a disastrous sea level rise is promoted by the Intergovernmental Panel on Climate Change (IPCC) as a part of global warning due to anthropogenic CO2 increase in the atmosphere.
19. There are physical frames and boundary conditions established by firm observations, which invalidate much of the scaring flooding concept of IPCC. See for example: Energy & Environment, 24, 509-536, 2013; Global Perspectives Geography, 2, 16-21, 2014; Coordinates Magazine, X:10, 15-21, 2014.

Toward an Understanding of Raw Emotional Feelings: The Source of Primal Affective Consciousness in Animals and Humans?

Jaak Panksepp

Jaak Panksepp is Professor of Neuroscience and the Baily Endowed Chair of Animal Well Being Science at the College of Veterinary Medicine of Washington State University (Pullman, WA). His research during the past half a century has focused on how animal brains control primal emotional behaviors and associated feeling states (as monitored by the rewarding and punishing properties of activation of specific emotional networks). The work has implications for understanding various human psychiatric disorders with novel ideas for development of new treatments of emotional problems in humans and other animals. It also has yielded new perspectives concerning the emergence of consciousness in brain evolution, which arises from deep subcortical brain regions, as do all the primal emotional feeling-systems.

One of the most important and most neglected areas of brain science has been the study of emotional feelings in other animals, as perhaps the only way to understand the neural nature of our own feelings. During the 20th century, many investigators of animal behaviors (especially the powerful "never mind" behaviorists) deem such experimental projects to be fool's errands, since there are supposedly no empirical ways to know whether other animals have any experiences. Indeed, the neuroscientific understanding of how brain activities generate affective experiences in humans is hampered by seemingly insurmountable empirical difficulties if we do not have adequate animal models. Thus, considering the shared evolutionary history of all mammals, our working premise has been that the study of raw, valenced emotional experiences in other animals -- namely, their primal positive and negative affective states (as monitored with the rewarding and punishing properties of deep brain stimulation

(DBS) -- is an optimal empirical gateway to understanding many of the neural details of the corresponding (evolutionarily homologous) processes in human beings. This is neither a common nor especially popular view in the rapidly growing abundance of higher-order neuro-psychological studies of human brains and minds.

In short, feelings are useful. They help us survive. If they can't help us survive, then by evolutionary theory they wouldn't be there. That people still debate, "Do other animals have feelings?" is surprising, because feelings are a very primitive form of survival. I have observed and documented the emotions on display in hundreds of experiments with chickens, rats, guinea pigs, cats, dogs, and other animals.

The main goal of my research across the past half-century has been to penetrate the nature of *primary-process* (i.e. "instinctual" or "unconditioned") human emotional feelings by studying evolutionarily related processes in animal models, where the necessary detailed neuroscientific work can be done. The pursuit of *causal* (also known as *constitutive*) brain inquiries remains ethically problematic in humans. A reasonable premise is that the neuroscientific study of the fundamental emotional "value systems" of animal brains will tell us more about the evolutionary foundations of our own emotional nature than any other available strategy. The weight of evidence currently indicates that a fundamental understanding of our primal affects—from RAGE (anger) to PANIC (extreme, precipitous sadness)--can be pursued rigorously in animal models, yielding abundant insights into our own inborn emotional qualities. Although there remain many historical and philosophical resistances to the acceptance of such strategies, the evidence clearly indicates that all mammals studied so far share ancient "instinctual" emotional brain networks as evolutionary birthrights: A key finding is that the artificial activation of specific brain regions, with precisely localized electrodes, whereby weak pulses of applied electricity can generate emotional action patterns. These evoked emotional states are 'desirable' (rewarding) and 'undesirable' (aversive/punishing) to both humans and other animals.

The feeling (affective) components of such brain emotional arousals presumably signal key survival issues, with all positive (rewarding/

desirable) feelings signaling probable paths of survival, and all negative (aversive/punishing) ones, the paths of destruction. This highlights how animal brain research finally allows us to scientifically penetrate the fundamental neural nature of shifting affective states of mind in humans through the close study of evolutionarily related brain processes in other animals. Indeed, all such emotion-evoking sites exist in deep sub-cortical brain regions, which are shared homologously (i.e., they are evolutionarily related) in all mammals. Indeed, at present, we have learned more about the subcortical functions of our brains by studying corresponding neural circuits of animal brains than we have from human research. This kind of scientific knowledge is invaluable for understanding not only the sources of affective mental states in animals, but also within our own brains. Of what greater use might such knowledge be? It can help us understand the affective nature of psychiatric disorders in humans, which provides critical neural knowledge for the development of new psychiatric medicines, especially for pervasive mood disorders such as depression and mania, and increasingly common childhood disorders, especially Attention Deficit Hyperactivity Disorders (ADHD) and autism.

Thus, my work has been among the first to explicitly address the sources of affective consciousness in mammalian brains—the various ways our brains allow us to experience diverse positive ("good"-desirable) emotions as well as various negative ("bad"-aversive) ones. Indeed, beside the various *emotional* feelings, such approaches may eventually also clarify the nature of *homeostatic* affects (bodily feelings such as *hunger* and *thirst*), and *sensory* affects (the pleasures and aversions of taste, smell, touch).

This strategy is not widely accepted yet because of historical forces that have molded the field of "*behavioral neuroscience*" since its inception about half a century ago. The supposition that internal feelings of other animals ("affective consciousness") cannot be studied in animals remains a pervasive scientific bias in the field of animal brain-behavior investigations. Rationales for such faulty beliefs include long-standing philosophical traditions such as "dualism"—that mind and matter are distinct types of entities in the world. And even those who do not succumb to such nonscientific thinking often assert that the empirical access to mental states can only be achieved through study of the only species that can speak about their mental states, namely we humans. Supposedly, the study of the brains and behaviors of non-speaking animals can never suffice.

Indeed, ever since Rene Descartes' famous treatise on *The Passions of the Soul*, this kind of thinking has led many scientists, starting with philosophical-psychologist William James (1842 –1910), to situate the 'locus of control' for our many affective (i.e., "valuative"), good/desirable and bad/undesirable feelings, within the highest regions of human brains, namely our expansive neocortices. If that were true, it would be most unlikely that definitive and useful knowledge about our emotional experiences could arise from the study of the 'less-encephalized," smaller-brained creatures that cannot talk. Thus, in the modern era, most neuroscientists accept that while emotional behaviors can certainly be studied in other animals, their affective experiences remain impenetrable —they cannot be empirically fathomed. But they are wrong.

As one influential investigator of emotional learning in rats opined in recent years: "We will never know what an animal feels." (*see Endnote 1*). Regrettably, such opinion leaders are simply sharing personal biases as opposed to any critical evaluation of the robust fact that wherever we arouse visually evident emotional "instinctual" action patterns with minuscule amounts of electricity—namely with DBS--the animals always treat the evoked internal states as "rewarding" or "punishing." *Rewards* and *punishments* are commonly used generic descriptors of the various objects and events of the world, from tasty treats to painful foot shock, that behaviorists have traditionally used to train animals in an enormous variety of learning tasks (since most behavioral neuroscientists are more interested in how memories rather than minds are created within brain dynamics.)

Of course, the early behaviorists were right that, before neuroscience, it was just a matter of opinion whether animals experienced their *rewards* and *punishments* as human beings most certainly do. But now, because of the kinds of neuro-scientific inquiries summarized here, a cogent interpretation is that other animals also experience their emotional arousals, which allows them to have useful memories that facilitate life-supporting choices. Regrettably, when many animal behaviorists flooded into neuroscience (when government grant support for their "never mind" animal research projects evaporated in the early 1970s), they chose to abide by their long-standing beliefs and biases about the scientific impenetrability of mental states in animals. Thus, rather than participating in the reasoned evaluation and discussion of the DBS evidence for "rewarding" and "punishing" brain systems, along with the many new human (and psychiatric) predictions that affective neuroscience offered in abundance, the discussion of such critical topics remained widely ignored in academic psychology. That bias has now dramatically delayed the development of new pre-clinical (animal-research based), knowledge about raw human feelings as well as the discovery of new treatments for psychiatric research. In any case, I persisted with that "unwelcome task," partly because of my own unusual personal and intellectual history.

At the beginning of my career, what convinced me that animal brain research could provide critical information to resolve the endless debates of the past, both historical and modern, on this topic of universal interest, was an evolutionary view of organisms that goes back to Charles Darwin, along with abundant neuroscientific evidence on emotions that goes back to the work of Walter Hess (at University of Zurich). Hess (1881–1973) first demonstrated that we could evoke one dramatic type of emotional display in cats, namely anger and subsequently fear by others, (and then all mammals that were studied, including humans) simply by stimulating very ancient regions of the brain, especially the hypothalamus. Hess's discovery was completely a matter of serendipity, since his main project was to simply map the "autonomic nervous system" within the nervous system with DBS in awake cats.

Hess's discovery of brain areas that modify our visceral activities, especially heart rate and blood pressure changes, brought him well-deserved Nobel Prize recognition in 1949 (the year our family arrived in America from Estonia, via war-torn Germany after the travails of WWII). Using direct electrical stimulation of specific brain regions, Hess identified many subcortical brain sites, especially in the hypothalamus, deep below the neocortex, where emotional behaviors were triggered in association with those autonomic arousals. However, Hess was very hesitant to ascribe emotional feelings to animals, largely because that was *not* his main interest. He also saw no reason to take a position on that contentious issue, especially since he suspected that his main-line of work might then be marginalized by the Anglo-American behaviorists who claimed that any discussion of mental processes in animals could be no more than soft-headed opinion (aka, scientific nonsense); so he chose to call it "sham rage," a mistake that he later regretted.

My Personal Entry into the Field

So how did I end up being the guy that challenged the behaviorist "never mind" biases in the field with piles of evidence and argumentation? There was little in my educational trajectory that could have suggested a career path toward working on the fundamental emotional circuits of mammalian brains. I started my undergraduate work in electrical engineering at the University of Pittsburg, but after changing majors several times, ended with a degree in psychology, and entered graduate work in 1965 at the University of Massachusetts in the specialization of Clinical Psychology.

A critical event led me to take that dramatic shift from engineering and basic sciences to an interest in how emotions are constructed in human brains: A summer job I had during the last years of college was as a night orderly in the psychiatric ward of a Pittsburgh hospital. There, for the first time, I was exposed (at least professionally) to the drama of human emotions that had gone astray. I began to be entranced by the scientific question "What is an emotional feeling?"

The ward was organized as one long linear space with three areas. The front or "easy" area consisted of living space for reasonably well-regulated, often short-term, patients. That was followed by a more regulated area for the more difficult-to-manage and the chronic cases. Then came a locked "padded-ward" of small individual rooms for the very hard-to-manage—the floridly psychotic, and all too often, violent individuals. After gaining some experience among the easiest patients, I worked in all the areas, and the evening shift allowed me abundant free time not only to get to know many patients, but also to read about their life histories and to see how they responded to the many psychiatric medicines that were coming into use in the middle part of the past century.

Of course no normal human being is a stranger to the power of emotional feelings in their own lives, and mine were surely as 'rich' as most. No doubt I wanted to understand the nature of human emotional feelings partly because I had had diverse emotional upheavals in my own life. Among the first was when I was less than a year old, toward the end of World War II, as my family was escaping from the advancing Red Army that swept devastatingly over Estonia and my parents' farm (on the outskirts of Tartu, Estonia). Indeed, during our escape, I almost died from a war-related injury. Of course, the lives of all immigrant families in the midst of that tragic war were full of emotional turmoil. So I was surely tutored in emotional experiences from my earliest years, which has motivated and perhaps informed my half-century of scientific research and conceptual reflections on the topic (see endnote 3). However, my scientific focus came to be riveted on how such powers of mind could ever be understood with scientific rigor.

I naively thought graduate studies in Clinical Psychology would introduce me to the scientific understanding of human emotional processes and their role in understanding and treating psychiatric disorders. I was surprised to learn that there was little neuroscientifically-based talk about the nature of emotional feelings in academic psychology during that behavioristic era, even though many speculative views had been advanced by diverse scholars from Descartes to Darwin, and onward. I shifted promptly during my first year of graduate work to the specialization of "physiological psychology" (or "behavioral neuroscience" as it soon came to be known in the late 1970s). That was the only place to get the needed neuroscience training to pursue my own fledgling inquiries into the nature of human emotions, namely by studying the underlying brain processes in animal models.

Little did I realize that it would be a long time--nearly till the end of my scientific career—before any empirically grounded discussion of the emotional experiences of other animals would emerge by others, at least within the rigorous neurosciences. After several years of intense work on brain self-stimulation systems during my first years of graduate work—

namely, on the so-called "brain reward system" that had recently been discovered by James Olds and Peter Milner (1954) at McGill University — I completed my PhD research mapping how DBS of specific brain areas, mostly in hypothalamic regions can instigate aggressive behaviors in rats. I was following not only the lead of Walter Hess, but John Flynn's seminal work at Yale on cats. In addition to characterizing two types of aggression that could be evoked by subcortical DBS—namely *emotional* or *affective* attack as well as *quiet-biting* attack (the former was accompanied by hissing and executed with ballistic intensity like intense human anger, while the latter was more pre-mediated, more predatory) —I decided to go one step further: I proceeded to determine whether the animal experienced anything by simply inquiring whether they would choose to turn those kinds of DBS on or off when given a chance by pressing a lever. Indeed, all animals exhibiting *affective* (seemingly "angry") attack, given the chance, would turn the brain stimulation off. In contrast, those exhibiting predatory type of attack consistently found the evoked states rewarding, since they all would willingly turn the stimulation on.

Clearly, animals did not just have emotional behavior patterns built into their brains, they also clearly had associated emotional feelings, at least to the extent that *rewarding* and *punishing* properties of those evoked emotional states could inform us about what was experientially happening in their "minds," a finding that has been validated through corresponding human studies. However, now some of the neurochemistries have been worked out (e.g., the neuropeptide Substance P promotes angry aggression), although none has yet been used to control pathological anger in humans.

Indeed, here was a robust and lawful relationship: Wherever in the brain we could evoke coherent emotional response patterns with DBS, those states were always either rewarding (SEEKING, LUST, CARE and PLAY urges) or punishing (RAGE, FEAR and PANIC emotions); for capsule descriptions of these systems, see below. This discovery remains the most compelling evidence for ascribing emotional feelings to other animals (which does predict human emotional feelings evoked by DBS of these same brain regions). Thus, it provides a scientific rationale for understanding the neuroanatomical and neurochemical nature of human emotional feelings by studying the ancient subcortical circuits of other animals.

Those diverse lines of evidence that finally allowed us access to the emotional feelings of other animals led to my 1998 monograph, *Affective Neuroscience* (Oxford Univ. Press), inaugurating a new field of brain science that sought to integrate our ever increasing understanding of how the neuroanatomies and neurochemistries of emotional brain systems of humans could be illuminated (with arguments conveyed in more popular and clinically relevant form in *The Archaeology of Mind* (Panksepp & Biven, 2012, Norton). Some critics have claimed that my approach is unacceptable because I have been ascribing human-type emotional feelings to animals (the sin of "*anthropomorphism*"— ascribing human feelings to animals, see Note 1), but that is not the case. What I have actually been doing is "*zoomorphism*" (trying to give primary-process animal feelings back to humans by understanding brain systems we share with all the other mammals.)

In any event, the analysis of brain emotional systems with DBS and evaluation of the *rewarding* and/or *punishing* properties of those evoked brain-mind states has been a workable project for some time—namely ever since Jim Olds and Peter Milner found that DBS of certain septal and hypothalamic region could produce rewarding effects back in 1954, while others concurrently showed that DBS of nearby brain systems were punishing (aka, aversive). What has been missing was a rich conversation, at least within traditional *behavioral neuroscience,* about the psychological nature of these shifting brain states, and their importance for understanding psychiatric disorders and thereby development of new treatments).

The conjunction of such behavioral and affective analyzes is especially important for three reasons: i) It allows us to have confidence that there are valenced feelings (positive/rewarding and aversive/ punishing states) that accompany DBS-induced emotional-behavioral arousals in other animals, just as in humans. ii) With advances in our understanding of specific brain circuits, it allows us to understand the most important neurochemicals that control emotional arousals in mammalian brains, including humans. iii) Thereby it provides new ideas for how to regulate emotional feelings, which may allow us to develop new psychiatric treatments through the development of new medicines as well as other more direct interventions such as DBS pacemakers placed into emotional systems that are imbalanced (*see endnote 3*).

To this day, DBS of subcortical brain circuits provides the best way to decode the neurology of emotional feelings, although newer, more refined techniques to achieve this have been developed—e.g., instead of using localized electrical stimulation of the brain, investigators are beginning to use DBS with light (via optic fibers implanted into specific brain regions) in appropriately genetically-modified animals, a technique called *optogenetics* (for recent use in animal emotional research, see endnote 4), and other modern techniques that I won't focus on here. These are bound to ratchet our understanding to new levels of detailed sophistication. However, it is unlikely that the key general principle will change.

The challenge now is to empirically distinguish the many types of positive and negative affective states brains can elaborate. We need to distinguish emotional affects from other types such as bodily homeostatic ones such as HUNGER and THIRST, and diverse sensory ones such as PAINs and PLEASUREs. Why? We are more likely to make major biomedical progress on major health issues, such as our current massive epidemic of obesity by development of appetite/weight control agents that truly reduce HUNGER feelings, etc., as opposed to simply those that reduce food intake in animals because the drugs made them feel bad/ill, etc.

The Diversity of DBS-Evoked Emotional Behaviors and Affective States in Animals

Why did I capitalize the emotional-system labels used above? It was simply because we need a special nomenclature for the many intrinsic affective (automatic, valuative) systems of animal brain. Here I just focus on the 'emotional' ones that are most important for a deep science of brain systems most relevant for psychiatry. As it stands, we can provoke at least seven types of emotional tendencies in animals with DBS, and I have chosen to call these fundamental emotions with specialized (but understandable) formal scientific names—SEEKING, RAGE, FEAR, LUST, CARE, PANIC, and PLAY. Indeed, they are sequenced in this way because of inferred evolutionary progressions, with SEEKING being perhaps the oldest and biggest emotional system of them all. We know these systems are built into the brain for one simple reason: The DBS consists only of unpatterned electricity applied to discrete loci in the brain--the simulation does not contain explicitly resolved "information" but only "energy". If you would stimulate the "innards" of your computer that way, you would ruin it. However, we can apply such electrical "garbage" into specific brain areas, and evoke coherent behaviors in animals, as well as, humans. If we were silly enough to do that to central processing units of a modern computer, the unforgiving law of "garbage in, garbage out" would prevail. But surprisingly, done within the right subcortical areas of the brain, the DBS is sufficient to trigger a variety of coherent behavioral and psychological "state shifts." Thus, within the brain, as we put "garbage" (or raw electrical energy) in, we get "coherence out" which means we are triggering pre-existing (i.e, evolved) coherence-generating mechanisms of the brain. And we can be confident that there are feelings involved in humans, because they can tell us so, but also in animals, because these brain stimulations are routinely *rewarding* (i.e., the SEEKING, LUST, CARE and PLAY systems), while others are *punishing* (i.e., the RAGE, FEAR and PANIC or "separation distress/sadness" systems).

Of course, we must also eventually "ask" animals whether the *rewarding* states are truly accompanied by different types of positive feelings, and whether the *punishing* states reflect different types of negative feelings. This remains to be well studied, but distinct feelings are commonly seen in humans stimulated in these brain regions (see endnote 5). One of the few empirical ways we could determine in animals whether the various rewarding and punishing feelings are distinct is by asking if they can discriminate between various rewarding brain sites, and between various punishing sites (when provided as cues for discriminative behavioral choices, such as whether or not to press a lever for food when hungry, or to avoid shock), but very few studies of that type have been conducted so far.

There is much work like that left to be done, but surprisingly few are pursuing such work, perhaps because necessary funding remains scarce for anyone claiming that this work allows us to understand certain human feelings by understanding the affective feelings of other animals. My emotion work was never explicitly funded by any governmental research agency. But when it comes to the issue of whether the knowledge we have obtained is promoting development of new psychiatric treatments, it is clearly well on its way (see Note 3). I will briefly highlight those achievements after summarizing the seven primary emotional systems that have been revealed with DBS. These systems are all situated subcortically, in brain regions where evolutionary homologies abound across all mammalian species. These systems consist of large transverse networks that interconnect brainstem circuits concentrated in midbrain regions such as the periaqueductal gray and ventral tegmental area, with various basal ganglia nuclei, such as amygdala and nucleus accumbens as well as cingulate and medial frontal cortex, via pathways that run through the hypothalamus and thalamus. Each of these *primary-process* emotional systems has abundant descending and ascending components that work together in a coordinate fashion to generate various instinctual (*primary-process*) emotional behaviors as well as the raw feelings normally associated with those behaviors.

These primal systems facilitate learning and memory formation by currently poorly understood "Laws of Affect" (which are *secondary-process*, namely unconscious neural ways to distribute affective survival-issues so they are integrated with outside world events, through learning and memory.) These lower systems control much of what happens within our diverse higher brain-mind mechanisms (what we call *tertiary-process* mechanisms, which become enriched in our neocortices, where specific cognitive tendencies are programmed largely by life experiences). Our suspicion is that many of those higher brain functions will never be well understood, namely neuroscientifically, until we understand the foundational affective processes. For instance, it is likely that learning and thinking may be heavily influenced by affective arousals.

Obviously there is much work left to do before we understand the primary (evolutionarily-provided or instinctual-*unconditional*) systems in detail, but we at least now know that all mammals have a variety of basic emotional systems that coordinate the autonomic, behavioral and raw feeling aspects of emotions. To understand the whole evolutionary and developmental complexity of the mind, we need to understand how these various levels of control operate within various poorly understood nested-hierarchies. In early infant development the “bottom-up” controls prevail, which with maturation allow “top-down” controls. The psychological features of these are very hard to decode in animals so far, and each level of analysis needs a distinct nomenclature.

This is not the place to elaborate on the many issues that affective neuroscientists still need to clarify. But, it is surely good to be conversant with the primal emotional networks that *Mother Nature* constructed in ancient regions of our brains long before *homo sapiens* conquered the world, and to recognize that these systems generate affective experiences, not just emotional behaviors. What is the function of the feelings? Presumably to unconditionally forecast survival trajectories, which need to be solidified by learning. Thus, we have to understand other organisms not simply through the study of their external behaviors but also to generate internal perspectives on their mental (experiential) lives. This is ultimately how my work contributes to our ongoing aspiration to understand the deep animalian nature that undergirds our human nature. How these affective systems actually interact with higher brain processes is an even more difficult work in progress. However, to flesh out the basics, here I briefly describe the seven *primary-process* systems that we’ve mapped that currently are supported by abundant empirical evidence from animal brain research (for more detailed coverage, see *Affective Neuroscience* (Oxford Univ. Press, 1998) and *Archaeology of Mind* (Norton, 2012)):

1) **The SEEKING/Desire System.** This is a general-purpose appetitive motivational system that may be essential for many other emotional systems to operate effectively. It seems to be a major source of the life-energy that has at times been called “libido." It induces animals to be intensely interested in exploring their world (with enthusiastic

curiosity) and with automatic learning, which gradually leads them to become excited when they are about to get what they desire. It eventually allows animals to find and eagerly anticipate all kinds of resources they need for survival, including water, food, warmth or coolness depending on the status of thermoregulatory systems, and their ultimate evolutionary survival need, sex. When fully aroused, the SEEKING urge fills the mind with enthusiasm/interest and motivates organisms to move their bodies seemingly effortlessly in search of the things they need, crave, and desire. In humans, this system generates and sustains curiosity from the mundane to our highest intellectual pursuits. When this system becomes under-active for various reasons, such as drug withdrawal, chronic helplessness/stress or neural deficits of old age, a form of depression results. When the system becomes spontaneously overactive, which can happen as a result of various drugs, behavior can become excessive and stereotyped, guided often by psychotic delusions and manic thoughts.

Neuroanatomically, the SEEKING system corresponds to the major self-stimulation system that runs from midbrain up to the medial frontal cortex. Animals will activate this system readily, whether with drugs of abuse such as cocaine or direct electrical or chemical stimulation of the brain. This system has long been inadequately conceptualized as a "reward, pleasure or reinforcement system." However from a more sophisticated neuropsychological perspective, it appears to be a general-purpose neuronal system that coaxes animals and humans to move energetically from where they are presently situated to the places where they can find and consume resources needed for survival. It permits learning by readily assimilating predictive reward relationships in the world. A critically important chemical in this system is dopamine. Dopamine circuits can energize and coordinate the functions of many higher brain areas that mediate planning and foresight and that promote normal states of anticipatory eagerness and apparent purpose in both humans and animals. Of course dopamine controls many other brain functions, from our motor actions, sense of time, learning and memory, to our enthusiasm about anything.

2) **The RAGE/Anger System.** Working in opposition to SEEKING is a system that mediates anger. RAGE is aroused by frustration and attempts

to curtail an animal's freedom of action. It has long been known that one can enrage both animals and humans by stimulating very specific circuits of the brain, ones that lie rather close to the trajectory of the FEAR system discussed next. The RAGE system invigorates aggressive behaviors when animals are irritated or restrained and also helps animals defend themselves by arousing fear in their opponents. Human anger may get much of its psychic energy from the arousal of this brain system; there are a number of well-documented cases where humans stimulated in these brain regions have exhibited sudden, intense anger attacks, with no external provocation. Key chemistries in this system are the neuropeptide, Substance P, and glutamate, which activate the system, and endogenous opioids, which inhibit the system, but these chemistries (especially glutamate and opioids) also participate in many other emotional responses as well as learning and memory. Specific medicines to control angry behavior in humans and animals could presumably be developed through further detailed understanding of this brain circuitry.

3) **The FEAR/Anxiety System.** A coherently operating FEAR circuit was designed during brain evolution to help animals reduce pain and the possibility of destruction. When stimulated intensely, the circuit leads animals to flee as if they are extremely scared. With much weaker stimulation, animals exhibit a freezing response, which is also common when animals are placed back into an environment in which they have been hurt or frightened. Humans stimulated in these same brain regions report being engulfed by an intense free-floating anxiety that appears to have no environmental cause. Key chemistries that regulate this system are Neuropeptide Y and Corticotrophin Releasing Factor (CRF); specific anti-anxiety agents such as the benzodiazepines inhibit this system (the first being the blockbuster tranquilizers Librium and Valium).

4) **The LUST/Sexual Systems.** Sexual urges are mediated by specific brain circuits and chemistries that are somewhat overlapping, but also quite distinct for males and females. They are aroused by male and female sex hormones, which control many brain chemistries including two neuropeptides whose synthesis is strongly controlled by sex hormones: Oxytocin transmission is promoted by estrogen in females and vasopressin transmission by testosterone in males. These brain chemistries help create

gender specific sexual tendencies. Oxytocin promotes sexual readiness and acceptance postures in females and vasopressin promotes assertiveness, and perhaps jealous behaviors, in males. Distinct male and female sexual circuits are constructed very early in life, and are activated by maturation of gonadal hormones at puberty. Because brain and bodily sex characteristics are independently organized, it is possible for animals that are externally male to have female specific sexual urges and, likewise, for some to be female in external appearance but to have male sexual urges. Some of the chemistries of sexuality, for instance oxytocin, have been re-used in brain-mind evolution to mediate maternal care—nurturance and social bonding—suggesting there is an intimate relationship between female sexual rewards and maternal motivations.

5) **The CARE/Maternal Nurturance system.** Brain evolution has provided safeguards to assure that parents (usually the mother) take care of offspring. The massive hormonal changes at the end of pregnancy (declining progesterone, and increasing estrogen, prolactin, and oxytocin) set the stage for the activation of maternal urges a few days before the young are born. This symphony of hormonal and neurochemical changes, especially the heightened secretions of oxytocin and prolactin, facilitate maternal moods which assure strong social bonding with the offspring. Similar neurochemicals, especially oxytocin and endogenous opioids, promote infant bonding to the mother. These changes are foundational for one variant of love.

6) **The PANIC/Separation Distress System.** All young mammals are dependent on parental care, especially maternal care, for survival. Young animals have a powerful emotional system to indicate they are in need of care, as reflected in their intense crying when left in strange places by themselves. These separation calls alert caretakers, mothers typically, to seek out, retrieve, and attend to the needs of the offspring. The separation distress system has now been mapped in several species; it is powerfully inhibited by endogenous opioids, oxytocin, and prolactin—the major social-attachment, social-bonding chemistries of the mammalian brain. These basic separation-distress circuits are also aroused during human sadness, which is accompanied by low brain opioid activity. Sudden

arousal of this system in humans may contribute to the psychiatric disorder known as "panic attacks."

What we learned from some of our earliest work on bonding, measured by separation-distress or PANIC calls, CARE is backed by brain opiates. We administered tiny doses of opiates to dogs, guinea pigs, and chickens. We found that young animals given tiny doses of opiates like morphine cried less or not at all when separated from their mothers. The same could not be said of anti-anxiety medications, suggesting anxiety grows more from FEAR circuits than the PANIC circuits involved in social separation, loneliness, and sadness. That appears to be one of the main sources of opiate addiction. People that are isolated, may suffer a lot of psychological pain and learn to treat themselves. Often they are self-medicating to ward of depressive feelings.

7) The PLAY/Rough-and-Rumble, Physical Social-Engagement System. Young animals have strong urges for physical play. This takes the form of pouncing on each other, chasing and wrestling. These actions can seem outwardly aggressive but they are accompanied by positive affect—an intense social joy. During these activities, rats make abundant high frequency (~50 kHz) chirping sounds that have many features resembling human laughter. It is interesting to note that there seem to be similarities between the subcortical brain circuits that mediate human laughter and play-induced chirping in rats. The most powerful evidence for an evolutionary relationship between positive affect and chirping is the fact that, if humans tickle rats, these vocalizations often go up to maximal levels, and the young animals rapidly return to the individual human and attempt to solicit more tickling (see http://www.wired.com/2013/09/tickling-rats-for-science/ and this video: http://tinyurl.com/ratslaughter). In contrast to positive affect, when negative feelings are aroused, animals commonly exhibit 22-kHz "complaint" type vocalizations and play temporarily ceases. A key function of the social play system is to facilitate the natural emergence of social dominance, negotiate rules and limits, and establish group dynamics. Play helps young animals to acquire more subtle social interactions that are not genetically-coded into the brain but must be learned. The play is one of the major emotional forces that promotes the epigenetic construction of higher social brains. This system, like all the other emotional systems, is concentrated in specific subcortical regions of the brain.

Video: NPR: Jaak Panksepp on Play and the Development of Social Skills - http://tinyurl.com/playsocialskills

The RAGE system helps animals obtain resources and get their way, but it also feels bad, as do the arousals of FEAR and PANIC, the two other brain states animals turn off if given the chance. The persistence of SEEKING is an indication that evolution itself, the driving force behind these emotional innovations, has built optimism into the brain. The positive emotions carry life forward. It's the negative emotions that have to deal with crises. So our basic nature biologically is positive. Think about it. If the negative emotions were persistently prevailing, we'd be wretched creatures from the outset, as is common in depression.

TED talk: The science of emotions: Jaak Panksepp at TEDxRainier - tinyurl.com/jaakted

As a result of our animal research, we have proposed that the failure of so many children, especially in one-child families living in big cities, to have sufficient free play, may be reflected in impulse control disorders such as ADHD, which may be gradually changing the fiber of our society. To try to counter such trends, we should consider building more play sanctuaries for our kids. We have done our share, but have failed to attract funding for the desperately needed studies in pre-school children. It is an issue that our society must attend to if we are going to promote optimal brain-mind maturation and reduced ADHD in our children. The power of PLAY remains poorly understood. But there are many mental-health implications of our ever-increasing understanding of the basic emotional systems of mammalian brains.

Psychiatric-Therapeutic and Philosophical Implications of this Work

As our research group clarified some of the details of the above emotional systems, the psychiatric implications became ever clearer. For instance: i) Autistic children may be socially aloof partly because they do not develop strong and specific social bonds, as are mediated by the PANIC, CARE and PLAY systems. ii) As already noted, ADHD reflects, in part, the fact that many young children are no longer getting adequate early social PLAY, which can in classroom situations be interpreted as an "impulse control disorder." Extra play may reduce ADHD symptoms in our children as it does in animals. iii) Finally, recently we have conceptualized how the affective underpinnings of depression may reflect excessive activity in brain PANIC systems and diminished activity of SEEKING and PLAY systems. Indeed, we have developed interventions to diminish activity of the first and to increase activity of the other two systems to counteract depression, with very promising initial human clinical results (for details, see Note 3).

Video: The Primal Power of Play - wsm.wsu.edu/s/we.php?id=352

Video: Jaak Panksepp on the brain and searching the web - tinyurl.com/webseeking

There are obviously many philosophical implications of this work. For me, the most intriguing is the neural nature of the primal animalian core-SELF-process—the coordination of psychomotor neurodynamics that help create the organismic coherence that is so evident not only in humans but in the actions of other animals with whom we are so fortunate to still share the earth. Some call this brain process "the soul." Perhaps those issues can be addressed empirically by the next generation of neuroscientists who take the inner life of other animals and their welfare, as opposed to just their behaviors, seriously. As a young student recently told me "other mammals are basically conscious in a way so similar to us that the situation demands we treat these animals far better than we do right now." Yes, but we also have to acknowledge that we have not yet learned to study how other animals think, although there are some visionary scientists who are making some progress on even those, more subtle, issues (see Note 6).

Endnotes

Note 1: There have been diverse (and mostly welcome) critics of my work, but none as persistent as those by Joseph LeDoux, who has asserted "We will never know what other animals feel." He has also claimed I have indulged in the cardinal scientific sin of *anthropomorphism*—the giving of human mental qualities to other animals (e.g., see p. 666 of his 2012 article on "Rethinking the emotional brain" in Neuron 2012 in *Neuron* (vol. 73, pp. 653-576). That accusation is not valid. In fact, I have been trying to understand if all mammals share very similar primal emotional processes, and since they do, I have been trying to return our shared emotional states back to the human species, which is formally called *zoomorphism*. The claim is simple: When we finally understand these cross-mammalian brain emotional systems, we will begin to finally understand such affective mental powers of our own brains, and thereby hopefully generate new ways of understanding various psychiatric disorders, and through that understanding, crafting new evidence-based medications and other treatments (see Note 3).

Note 2: For a more complete description of my life trajectory, see a comprehensive "Interview with Panksepp" in the *American Journal of Play* (Winter 2010 issue) - tinyurl.com/jopinterview and for extensive interview, see https://www.youtube.com/watch?v=u4ICY6-7hJo

Note 3: Autism issues, including first neurochemical theory of the disorder, are summarized in: Panksepp, J., A neurochemical theory of autism. *Trends in Neuroscience*, 1979, 2, 174-177. For ADHD relations to PLAY in: Panksepp, J., (2007). Can PLAY diminish and facilitate the construction of the social brain. *Journal of the Canadian Academy of Child and Adolescent Psychiatry,* 10: 57-66.

Depression therapeutics are summarized in: Panksepp, J., Wright, J.S., Döbrössy, M.D., Schlaepfer, T.E., & Coenen, V.A. (2014) Affective neuroscience strategies for understanding and treating depressions: from preclinical models to novel therapeutics. *Clinical Psychological Science*, 2, 472-494 doi: 10.1176/appi.neuropsych.11080180., as well as Panksepp, J. & Yovell, Y. (2014). Preclinical Modeling of Primal Emotional Affects (SEEKING, PANIC and PLAY): Gateways to the Development of New Treatments for Depression. *Psychopathology,* 47, 383-393. DOI: 10.1159/000366208 (the full report of the anti-suicide and anti-depressive effects of low doses of the "safe" opioid buprenorphine in humans is currently "in press" with the *American Journal of Psychiatry* (Yovell, et al., 2015).

Note 4: Lee, H., Kim, D.-W., Remedios, T.E., Chang, A., Madisen, L., Zeng, H. & Anderson, D.J. (2015) Scaleable control of mounting and attack by $Esrl^{+}$ neurons in the ventromedial hypothalamus. *Nature,* 509, 627-632.

Note 5: For a summary of early work see: Panksepp, J., (1985) Mood changes: In P.J. Vinken, G.W. Bruyn, & H.L. Klawans (Eds). *Handbook of Clinical Neurology* (Revised Series). Vol. 1. (45): *Clinical Neuropsychology*. Amsterdam: Elsevier Science Publishers, pp. 271-285. For more recent discussions see: Anderson, R.J., Frye, M,A,, Abulseoud, O.A., Lee, K.H., McGillivray, J.A., Berk, M., & Tye, S.J. (2012). Deep brain stimulation for treatment-resistant depression: efficacy, safety and mechanisms of action. 36,1920-33. and Rodríguez, R.L., Miller, K., Bowers, D., Crucian, G., Wint, D., Fernandez, H., Foote, K.D. & Okun, M.S. (2005). Mood and cognitive changes with deep brain stimulation. What we know and where we should go. *Minerva Medicine*. 96, 125-44.

Note 6: Among my favorites in this emerging genera of research are: Rygula, R., Pluta, H., & Popik, P. (2012). Laughing rats are optimistic. *PLoS ONE*, 7, e51959 and Steiner, A.P. & Redish, A.D, (2014). Behavioral and neurophysiological correlates of regret in rat decision-making on a neuroeconomic task. *Nature Neuroscience,* 17, 995-1002.

Is What We Own Who We Are?

James K. Beggan

James Beggan was born and raised in Buffalo, New York and started college the same year the notorious Blizzard of '77 struck Buffalo. It was this event that motivated him to go to graduate school at the University of California, Santa Barbara. He graduated with a Ph.D. in psychology in 1989 and was appointed an assistant professor of psychology at the University of Louisville. In 1995 he was promoted to associate professor. In 2002, he transferred from psychology to sociology and was promoted to full professor of sociology in 2012. Although his early work involved understanding psychological factors related to decision making, his more recent research reflects interests in the representation of gender in mass media, the nature of sexuality as presented in pornography, and the social psychological underpinnings of sexual behavior. In addition to working as a professor, he can be found swing dancing between two and three nights a week, and, on occasion, performing at open mic comedy nights.

Where do we get our research ideas?

Scientific dogma found in undergraduate textbooks says they come from the logical positivist method of testing theories with reference to the external world. Textbooks in research methods present the job of scientists as an orderly process that involves inventing theories that generate hypotheses. Those rationally derived hypotheses are subsequently tested with empirical research. When obtained data are consistent with the hypotheses, faith in those hypotheses and the underlying theories that generated them are strengthened. When the data disconfirms what is expected, faith in the theories is weakened. If the theory disappoints often enough, the diligent researcher, like a disillusioned lover, begins looking elsewhere.

At some level, I would endorse the soundness of the logical positivist algorithm. I was taught that this is how research is done when I was in graduate school, and I have taught a new generation of students the same mantra. There is only one problem: While it certainly sounds like an efficient way to do research, and perhaps other scientists work in this manner, my own efforts have almost never followed this yellow brick road to success. I've never awakened and said, "I think I'm going to disprove a theory today." Instead, my research ideas have come from the oddest of places: random statements from friends, truisms from teachers presented without research support, or convenient happenstance. Similarly, my choice of research methods has stemmed less from the demands of the research question and more from the social and economic realities of where I was doing my research. Instead of asking, "What do I need to test this idea?" I've asked, "Since this is what I have to work with, what I can I investigate with it?" You know that quote, "To someone with a hammer everything looks like a nail." But people who use it to disparage people with hammers looking for nails fail to recognize that some things are, in fact, nails and need to be hammered.

The scientific method tells you what to do once you've gotten the idea. But I've never seen an undergraduate textbook that told the reader how to create ideas in the first place. After being a professor for 24 years, a graduate student for 5 years before that, and an undergraduate for 5 additional years, I realize that I've never learned or talked about where famous people got their famous ideas. Instead, professors talk about the famous ideas, what research has been conducted to support or refute a famous idea, and the strengths and weaknesses of a famous idea. The only exceptions that I can think of are the few legends about great scientists: How Friedrich August Kekulé realized that benzene had a ring shape after dreaming about a snake biting its own tail, how Richard Feynman—the brilliant physicist—had an insight into a research problem while watching a plate spinning in the air, and how Albert Einstein had the inspiration for the theory of relativity by imagining what it would be like riding his bicycle on a beam of light. But as I mentally review all the ideas I have read about or studied or done research on in the social sciences, I realize I cannot think of a single instance where people have talked about from where their idea came. Maybe these stories exist somewhere in a collected volume, but I am unaware of it.

Why don't researchers dwell on where they got their ideas? Maybe because talking about those kinds of things is thought to get in the way of the transmission of scientific information. If I only have seven or eight precious journal pages, I can't afford to waste them explaining where my idea came from instead of what my idea is and why I think it's a correct one. Do people not talk about it because they are embarrassed to admit their ideas come from haphazard or banal sources? Given what psychology knows about how people process information and make decisions, is it also possible that people genuinely *don't know* from where their ideas originate. Maybe I am anomalous because I know where at least some of my ideas came from. Or maybe my only aberration is being willing to address this subject by admitting I have not done things the right way.

One important explanation for the path my research has taken is the influence of collaborators and co-authors, such as Scott Allison and Patricia Gagné, and my own professors in graduate school, David Messick and Charles McClintock. These people have helped me to make the ideas better than I would have done on my own. But recognizing the assistance I've gotten from productive colleagues doesn't explain where I got the idea in the first place. The particular reasons I've worked on the research I've worked on range from environmental, like where I lived or went to school, to social, like comments from friends and colleagues, to academic, like how my professional training subsequently influenced the way I looked at the social world.

These causal factors exerted influence even before I started college. My childhood goal was to be a writer (of science fiction, specifically, but back when science fiction was a maligned specialty field). I thought of becoming an English major but during freshmen orientation I discovered that I would need a second language (I had little talent for learning languages, a fact I realized in high school) and that made me switch on the spot to psychology. I reasoned my efforts in writing the great American novel would be aided if I had a deeper understanding of human behavior.

As a result of this snap decision as a 17 year old, my path toward an undergraduate degree in psychology from the State University of New

York at Buffalo (SUNY Buffalo) started. Without necessarily intending to do so, I gave myself an unofficial emphasis in social—as opposed to areas such as cognitive, developmental, or abnormal-- behavior. Given that starting point, when I decided to go to graduate school, it made sense to apply to psychology programs, specifically social psychology programs. Thus, it is not surprising that when I got my PhD in psychology (with a more official emphasis in social psychology), I applied for jobs that called for a social psychologist, for the simple reason that I wanted to maximize my likelihood of obtaining employment. Aside from my newly printed doctoral degree, my other major credential was that I had published research in journals with the words "social psychology" in the titles. I got a job as an assistant professor of psychology at the University of Louisville in 1989.

But now, 24 years later, I am a professor of sociology and my area of interest has shifted from what would normally be studied by a social psychologist to something simultaneously narrower than social psychology and also more surprising. If I had to say what my area of expertise is, I would say "sex." If I had to describe my dominant research method now, I would say qualitative methods, specifically text analysis.

How did I go from being a quantitative social psychologist who studied game theory to a qualitative sociologist who studies sex? Answering this question is the intention of this essay. The research I want to talk about ranges from how people maintain a positive sense of self to the way that sex roles are represented in sexually explicit material. The research methods I used included carefully controlled laboratory studies but also the content analysis of magazines.

The importance of what appears to be random but relevant processes began even earlier than when I first started conducting and writing up my own research. It was an important contributor to where I went to graduate school and what I did while I was there. When I was in graduate school, I did research situated in terms of problems involved in making rational decisions. The method of choice was the famous prisoners' dilemma game. This game illustrates the conflict that can occur between decisions made to further self-interest versus decisions intended to benefit the collective. The name comes from the scenario that was first used to explain this self vs. others conflict. Imagine that two criminals have been arrested on suspicion of bank robbery. The police cannot prove they committed the theft but since they were arrested with unlicensed handguns, they can be each sent to jail for 1 year for unlawful possession of a weapon. The police separate the two men and tell one of them that if he turns state's evidence on his collaborator, he will get off with no jail time for being cooperative but the other man will go to jail for the maximum of 7 years. The police also tell the other criminal the same thing. Each criminal knows that the other has the same offer. The two men have a conflict. If they stick together, there is only the handgun charge. They each go to jail for a year, not too bad in the big scheme of things. If they both confess, they will each go to jail for 5 years for admitting they robbed a bank. The biggest danger is that one will cooperate with the police and the other will not. In this case, the trusting one gets heavily punished and the betrayer gets completely off the hook. The conflict of the prisoner's dilemma stems from the conflict between loyalty to your partner and the fear of being taken advantage of by that same partner to whom you are being loyal. From a strictly logical point of view, each person should confess, hoping the other does not. But if each person thinks things through logically, they both end up confessing and worse off than if neither had confessed.

The prisoners' dilemma is often presented in a matrix form, which looks like this:

	Prisoner 2 Confesses	Prisoner 2 Does Not Confess
Prisoner 1 Confesses	5 years each	Prisoner 1 gets no jail time Prisoner 2 gets 7 years
Prisoner 1 Does Not Confess	Prisoner 1 gets 7 years Prisoner 2 gets no jail time	1 year each

Video: Rational Choice: The Prisoner's Dilemma - http://tinyurl.com/zbh9gjb

Literally thousands of studies have been conducted under the umbrella of "mixed motive games" in the hopes of better understanding how people resolve this conflict. My own contribution involved thinking about personality differences between those who view the prisoners' dilemma as a cooperative environment ("how can we both get ahead?") or a competitive environment ("how can I get ahead of my opponent?"). To examine how cooperators and noncooperators thought about the competition-cooperation dimension, I had them write cooperative and competitive behaviors that they thought were more strongly associated with the self or with others. I found that cooperators and noncooperators think differently about the meaning of cooperation and competition. Cooperators tend to view cooperation in terms of morality and see cooperation as good and competition as bad. Competitors, in contrast, think about the cooperation-competition dimension in terms of power. For them, cooperation is viewed as a sign of weakness and competition is viewed as a sign of strength. This discovery can help explain why trying to resolve conflict may be so difficult. Cooperators would assume everyone would want to cooperate (after all, who wants to think of themselves as "bad") but competitors eschew cooperation on the grounds that doing so would make them look ineffectual. If some people think in terms of competition as being strong and others think of it in terms of being bad, it is not surprising that they might disagree on how to resolve a conflict.

If you are interested in reading a copy of this paper, you can find out more at: http://psycnet.apa.org/journals/psp/55/4/606/.

Why did I get involved in this line of work? It was the interest area of my two major advisors in graduate school. Why did I end up with them as my advisors? Because when I applied to graduate school I applied all over the country, but the schools that expressed the most interest in me were the ones that had faculty with interests in common with my undergraduate honors thesis advisor (an excellent scholar named Dean Pruitt). Why was he my advisor? Because when I decided to get involved in the kind of independent studies that might help me get into graduate school, one of his graduate students had put up signs in the psychology department advertising for research assistants. I saw the sign, called the number, and started working in Dean Pruitt's lab.

When deciding where to go to graduate school, my final two choices were the University of Illinois at Urbana-Champaign and the University of California, Santa Barbara. A number of sources told me that the University of Illinois was the better school, but I wanted my graduate school years to be partly an education but also partly an adventure. Spending five years in the relatively small town of Urbana-Champaign seemed like a step down. When I visited the campus, one of the graduate students told me that the only thing to do on the weekends was to drive up and down the street in pick-up trucks. I visited in the winter. I stayed overnight with a graduate student and one of the strongest memories I had of the visit was being freezing cold all night long. I slept with my coat as an extra blanket and took a shower ridiculously early in the morning to try to get warm. Contrast that experience with the one UC Santa Barbara created. Although I never visited the campus before accepting their offer, during the admissions process, one dreary Buffalo winter day, I received a slim manila envelope in the mail. Inside was an 8 x 10 glossy aerial photograph of UCSB. The photograph made the campus look like a resort hotel on the French Riviera in the summer time. I still have the picture hanging on a bulletin board next to my office desk.

Without a doubt, Santa Barbara sounds attractive as a place to live. The biggest downside: The cost of housing. But how did living there influence my academic research career? I was a graduate student between the years 1984 and 1989. While I was making slow and steady incremental process in getting a PhD, all around me in southern California there was a tremendous housing boom. People were making thousands of dollars of profit in southern California purchasing a house in January and selling it in June.

At the same time I was working on my classes and research, I also worried about my future. Would I get my degree? If I did, would I be able to find a job? Would it be an academic job, or would I have to settle for industry? What if I couldn't get a job at a university or in industry? What would I do then?

I found some small comfort in the idea of getting a real estate license. With one of those, I would have a back-up career, if I needed one. And if I ended up as an assistant professor, buying and renting property seemed like an appealing way to make extra money. There were little independent schools in Santa Barbara that taught courses geared toward meeting the requirements for being a real estate agent or building contractor. I took one. I learned about mortgages, how to prepare real estate contracts, and the basics of real estate mathematics. Compared to graduate statistics and prepping for my PhD qualifying exam, the course was rather easy and almost relaxing.

The instructor for the class was a real estate agent who repeatedly talked about the psychological benefits of home ownership, like being a better neighbor or having greater civic responsibility. But when I started looking for research evidence, I found virtually nothing. At that time, I was too busy finishing my degree requirements for the PhD and applying for jobs to take a side-trip down the research path of the psychology of ownership.

My first (and only) job (thus far) was at the University of Louisville. My appointment was as an assistant professor of psychology. Due to circumstances too difficult to summarize in brief, I ended up not having the resources to do research in the spirit of game theory and group dynamics. As I searched for other possible research topics, I started to think more seriously about the psychology of ownership. I found little research with the exception of two other scholars: Lita Furby, who did research focused on how children thought about possessions, and Floyd Rudmin, who had published quite a bit about the topic.

My training at UC Santa Barbara had focused on game theory and especially on how people deviate from rational choice. I applied this orientation, what I had discovered from studying game theory, especially why people deviate from rational choice, to thinking about ownership. From this perspective, the simplest, most basic question that I could formulate was: Does being the owner of an object make people like it more? Robert Zajonc showed that both humans and animals will express a preference for a stimulus to which they have been previously exposed. He called this phenomenon the "mere exposure effect," and I came up with the not-very-original "mere ownership effect" for my hypothesized relationship between ownership and attraction.

In nosing around the behavioral economics literature, I had come upon the term "the endowment effect," which stated that people wanted to sell a possession at a higher price than potential buyers were willing to pay. This was a very similar, perhaps too similar, idea. How was the mere ownership effect different than the endowment effect? A difference emerged in terms of how the effect was explained. The endowment effect was explained in terms of the hugely influential prospect theory. I had first learned about prospect theory in graduate school. Although initially presented in a very mathematical form, the basic idea was simple to understand and had the added advantage of feeling intuitively correct: Losses loom larger than gains. In other words, losing $10 feels worse than gaining $10 feels good. With the example of ownership, according to prospect theory, selling something you own can be construed as a loss that hurts more than the gain of the object feels good to the buyer.

From my way of thinking, however, the mere ownership effect did not stem from the desire to avoid loss when a possession was sold. Instead, it originated in the idea that what we own can be considered an extension of who we are. This idea was not original with me. It can be traced back in psychology to the writings of William James, who said, "Between what a man calls *me* and what he calls *mine* the line is difficult to draw….a man…is the sum total of all that he CAN call his, not only his body and psychic powers, but his clothes and house, his spouse and children, his ancestors and friends, his reputation and works, his house and yard, and even his yacht and bank account. All these things give him the same emotions. If they grow and prosper, he feels triumphant; if they dwindle and die away, he feels cast down…." This idea was presented in a more contemporary manner by Russell Belk with regard to what he called the "extended self."

The way that William James presented it, the self responded to fluctuations in the value of possessions. If a possession were to increase in value, we might expect a corresponding increase in the perceived value of the self. But I wondered if it could work the other way: Would a person work to enhance the value of possessions as an indirect means of enhancing the self? In other words, rather than think of the enhancement effect as an outgrowth of an asymmetry in how people thought about gains and losses, what if we thought about it in terms of the motivation for self-protection? There was already a huge literature on the various ways people went about trying to protect themselves from psychological harm with regard to a threatened self-esteem. The mere ownership effect was one more area where these principles of self-enhancement could apply. Why not conceptualize the mere ownership effect in terms of the motivation to maintain a positive view of self?

To make a case for the mere ownership effect, I conducted several studies. In the first experiment, people were asked to evaluate a series of small objects that were described as possible gifts that companies might give out, such as a pen, mug, or beer cozy. Half the subjects were told that they were being given the beer cozy for their help in doing the research. The other half did not receive the ownership manipulation. Subjects then rated the objects along several dimensions. As expected, subjects in the ownership condition rated the beer cozy as more attractive than subjects who did not receive ownership.

A second study ruled out a rival explanation based on the length of time people spent looking at the object. In this study, the experimenter had each subject look at each object for the same amount of time. The purpose of this restriction was to avoid the possibility that people in the ownership condition might spend more time examining their owned object and that the length of exposure, rather than ownership, was producing the effect.

The third study tried to justify an interpretation in terms of self-maintenance as opposed to loss aversion. In the third experiment, I gave subjects rigged task performance feedback. They were given a series of very difficult verbal puzzles to solve. Some subjects were told that even though they solved very few of the puzzles that was okay because the problems were difficult. Other subjects were told that it was surprising that they had solved so few, given that most people solved many more problems. I found that people showed a greater mere ownership effect when they failed in comparison to when they succeeded. The finding that people compensated for a threat to the self by reaffirming the value of a possession provides some insight into how to think about the psychological nature of ownership. Possessions—even relatively minor ones—can be thought of as being incorporated into the self. Once an object becomes part of the extended self, how it is treated may have an effect on how the owner sees himself or herself. In a sense then who we are depends in part on what we own.

If you are interested in reading this paper, you can find out more at: http://psycnet.apa.org/journals/psp/62/2/229/.

Where did the idea come from for the mere ownership effect? One explanation is that it came from the repeated comments of the real estate instructor. Being in California had made me more aware of housing prices than I ever had been growing up in Buffalo, NY. I probably would not have taken a real estate class had I gone to graduate school in Illinois. A training in graduate school that focused on the role rationality had (or didn't have) in people's decision making encouraged me to think about how ownership could serve as the basis for systematic deviations from rational choice. Knowing about Zajonc's mere exposure effect made me think about looking for a minimal level of influence.

If you were to look at my publication record at the turn of the century, there is this unexpected shift from the careful reporting of laboratory research grounded in the concepts and methods of social psychology. Suddenly, there are papers on *Playboy* magazine.

The shift to *Playboy* started because of a random comment from a colleague. She made a statement about how *Playboy* objectified women. Without fully thinking through the possible politically incorrect implications for my views on gender politics, I said, "I don't know. I don't think *Playboy* is that bad..."

"How can you think that?" she asked.

"Well, the *Playboy* Advisor"—the question/answer feature of the magazine—"always seemed pretty reasonable to me." I shrugged off the conversation with a "maybe we should do research on that someday…."

This was before the Internet became the repository for just about everything written about anything. Now, you can go onto the web and buy access to searchable digital copies of the complete run of *Playboy* for a few dollars a month, but when I was doing my work, the hardest part of doing research on *Playboy* was actually getting copies, especially older issues, of the magazine. You had to go to used and rare bookstores, and getting copies of early issues was cost prohibitive on an associate professor's salary. As fate would have it, in Louisville there was a bookstore with a back room filled with old *Playboy* magazines. They were reasonably priced and the more you bought the cheaper they got per issue. In groups of 10 the price went down to a dollar per issue. And this store had dozens of them.

Buying in bulk allowed me to accumulate virtually a complete run of *Playboys*—and the all-important Advisor columns—from about 1965 to the present. Not really sure of my ultimate research goal, I began to read through the Advisor letters. Certain themes started to emerge. The letters covered a huge range of topics. Sex was a big one, of course, but there were also letters from married men asking questions about how to make their marriages better, questions on how to select wines, how to buy stocks, how to wear suits. There was significant interest in martinis—fueled by the popularity of James Bond—debating the advantages of shaking, rather than stirring, them. An excellent, and sometimes humorous, introduction to the contents of *Playboy* Advisor letters can be found in *Dear Playboy Advisor: Questions from Men and Women to the Advice Column of Playboy Magazine* by Chip Rowe.

Playboy billed itself as entertainment for men, but if that was the case, then why were there a significant number of letters from women? The feminists said *Playboy* was about men objectifying and subjugating women. If that was true, then why did the *Playboy* Advisor often take the woman's side? If *Playboy* readers were sexist men who didn't care about the feelings or opinions of women, then why did the *Playboy* Advisor even exist? People ask questions when they don't understand something. Clearly, it seemed men were trying to understand—not subjugate—women.

When I looked at previous research on *Playboy* published in psychology and sociology journals, I found that virtually all of it focused on analyzing the centerfold, usually with the goal of proving that *Playboy* Playmates were thinner than average women. To me, it seemed that scholars doing research on *Playboy* did exactly the same thing that they accused readers of doing: They only looked at the Playmate's pictures. They ignored the text that accompanied the pictures, and they ignored the millions of other words and images in the magazine. What would happen if researchers did what readers ironically said they did? That is, what if I really read *Playboy* for the articles?

If you actually read the letters like I did—all 7000 of them across 40 years of publication—you realized that the men who wrote letters to the Advisor cared what women thought about them, they cared about what other men thought about them. Letters were written out of a basic insecurity rather than some kind of stereotyped masculinity.

The basic idea of a paper began to emerge from the recognition that it seemed like the *Playboy* Advisor was trying to help men understand their social world: how to get along better with women, potential girlfriends and wives, co-workers, and even the waiter who poured the wine and could be intimidating by his knowledge of drink. The *Playboy* Advisor was a benevolent big brother, not the all-seeing, all-spying figure in the novel *1984*, but rather, an older, wiser, and stronger protector. The *Playboy* Advisor did not work to reinforce gender stereotypes. Instead, he worked to refute them, with copious examples of letters from people whose problems did not conform to what would be expected from gender stereotypes.

You can find out more about this paper at: http://men.sagepub.com/content/9/1/1.abstract.

After the initial paper was written, submitted, and accepted for publication, I remember waking up one morning with two different but compatible thoughts. One was now that I had a huge run of *Playboy*s in my office with probably over 80,000 pages and millions of words sitting there waiting to be examined, I should probably do something with them.

The second thought was, "Why a rabbit?" If *Playboy* was about promoting masculinity, even hypermasculinity, and defending misogyny, then why was the symbol of the magazine a white rabbit? What are the main characteristics of rabbits? Rabbits were prolific, true, but they were also cute, soft, cuddly, and prey rather than predator, hardly a fitting image for a magazine that had as a defining goal the reification of traditional masculinity.

The question that echoed in the back of my head was, "Why a rabbit?" I reasoned that the most salient aspect of the magazine was the cover. The cover was intended to grab the potential reader's eye and get him (or her) to fork over money.

People only familiar with more recent incarnations of *Playboy* won't appreciate the humor and art that went into the covers in the earlier years of the magazine. You can see some of the earlier covers at: http://www.pinterest.com/nichenesses/vintage-playboy-cover-nicheness/.

One convention that started with the second issue was to feature a rabbit image somewhere on the cover. The rabbit might be a cartoon figure on a date with a woman, or it might be as abstract as a curl of a woman's hair or the shimmer of a bathing cap reflected in a pool of water. Sometimes the rabbit image was easy to find, but in other months it was quite difficult. In fact, sometimes in the letter column, *Playboy* had to explain to frustrated readers where the rabbit was hiding.

One thing I noticed about the rabbit was that sometimes he owed his existence to the efforts of women on the cover. One month featured a woman cutting paper doll rabbits, another month showed a woman creating a shadow puppet with her hand. Sometimes the rabbit image was contained in the woman's own body. In one of the most famous covers, the cover model was lying down with her legs in the air. Her torso formed the rabbit's head, and her legs represented its ears.

I coupled this observation with Hugh Hefner's first issue editorial in December 1953, when he wrote, "We like our apartment. We enjoy mixing up cocktails and a hors d'oeuvre or two, putting a little mood music on the phonograph, and inviting in a female acquaintance for a quiet discussion on Picasso, Nietzsche, jazz, sex." I realized that *Playboy* wasn't about reifying traditional masculinity. Instead it was more about redefining masculinity to include a great many more traits, some of which were more typically associated with women than men. A man who liked his apartment and liked to cook and discuss art and philosophy seemed like an ideal boyfriend. He seemed too good to be true, in part because he almost sounded like a woman. Don't forget this was decades before invention of the "metrosexual," a term popularized by the book *The Metrosexual Guide To Style: A Handbook For The Modern Man* by Michael Flocker.

Why a rabbit, even though a rabbit had traits that could be considered more feminine than masculine in nature? A rabbit symbolized the merger of traditionally feminine traits (cute, soft, and cuddly) with traditional masculine traits (sexually prolific). The rabbit was most often represented as a successful businessman or executive, upper class but definitely masculine roles. The history of the famous Playboy trademark can be found at http://www.designboom.com/portrait/playboy.html.

In a complementary manner, when I looked at the text feature that accompanied the photographs, I realized that Playmates were not presented in a manner that confirmed stereotypes of femininity as much as it broadened the cultural view of femininity to include women who professed to possess a number of non-stereotypical attributes. The women in *Playboy* were presented as strong, confident, and ambitious. Most were willing to delay getting married and having children to pursue their careers. A lot of Playmates were in college or were college graduates. Some had advanced degrees. And, of course, the most obvious and non-stereotypical thing about them: they were willing to pose nude for a national magazine.

In brief, then, *Playboy* was less about maintaining distance between men and women than it was about creating a bridge between masculinity and femininity.

After 13 years in the psychology department, I moved to sociology. The psychology department had become more focused on brain science and, as a social psychologist, I found myself not fitting in very well. As a field, sociology seemed a better fit. The sociology department at the University of Louisville emphasized both quantitative and qualitative methods. My earlier work was quantitative in nature. The *Playboy*-related material definitely fit into the category of qualitative sociology.

Two facts pushed me further in the direction of making the leap from quantitative psychologist to qualitative sociologist. The first was that the sociology department did not have the traditional undergraduate psychology pool where students were compelled to participate as subjects in research projects sponsored by faculty and graduate students. As a result, it was difficult to collect the kind of data with which I was used to dealing.

The second factor was that the person in sociology in charge of putting together the course schedule asked me if I wanted to teach "Human Sexuality," given that I had published several papers on *Playboy*. In a backward-logic kind of way, I had become enough of a *de facto* expert on sex by the small virtue of having published papers on one very particular and very narrow aspect of human sexuality to justify teaching a survey course in the subject. I said "yes" partly because I thought it was amusing to think of myself teaching a course in human sexuality, given my starting point in something as dry as game theory. The idea of recreating myself in my new department by teaching very different courses on very different topics was appealing.

Of course, in order to prepare to teach the class, I had to do a lot of reading. I got an undergraduate text on human sexuality as a starting point, started skimming the abstracts of the *Journal of Sex Research*, and thought about topics on human sexuality that might inspire interesting lectures. Now that the sex class has become part of my teaching repertoire, I am more vigilant to articles I see in magazines and online that talk about sex in an offbeat way that might appeal to my students. Other people--past students, friends, and colleagues—send me links to articles I wouldn't have seen on my own.

The old joke is that men think about sex every ten seconds. Now I was in the position of thinking about sex in a more abstract way, as a sex researcher rather than merely as a consumer. The more I taught sex and thought about sex in a scientific manner, more ideas came to me that could be made to serve as the basis for research. I thought about aspects in human sexuality from the prisoners' dilemma perspective because that approach—part of my early training—really stuck with me. As an example, think about the difference in the way that men and women achieve and express orgasm. With men, the most obvious sign is ejaculation. The seminal fluid is "proof" that the male has ejaculated and has enjoyed himself. But what about women? For the most part, there is no obvious sign that a woman as achieved orgasm. How can you tell? Take a look at Meg Ryan's orgasm scene in the movie *When Harry Met Sally*: http://tinyurl.com/oqc5y2v

You have to look for indirect signs, such as moans and groans. Or ask her. But it is always possible to fake these indirect signs. Or she might just be flat-out lying. How can a woman "prove" she's had an orgasm? In class, this kind of question often leads to a discussion of how often or whether or not a woman should fake orgasm. Faking it is often viewed by the women in my class as a good thing because it makes men feel better about themselves and saves the woman an uncomfortable but frank conversation about her failure to fully enjoy the sexual experience. In other words, from game theory perspective, faking it represents a defecting choice. It is a short term gain (people are relieved of uncomfortable conversations or awkward realizations), but in the long term, the woman "teaches" her sex partner about what she needs to achieve orgasm. And the partner "learns." But what is being taught and what is being learned are false sets of information. The short term avoidance has led to a long term loss (a failure of sexual communication).

Another project represents a hybrid of my early research on ownership with my more recent work on human sexuality. I've applied the idea of self-enhancement to how people judge their own sexual competence. This project started out as a throw-away line in my human sexuality class: "Who here thinks they are better than average at....kissing?" What I discovered that virtually everyone thinks they are on average better than others as sex partners. But we know that not everyone can be better than average. I called this the "good-in-bed effect" and demonstrated it using several laboratory techniques. The implications of this phenomenon relate to marital satisfaction. An important source of marital discord reflects dissatisfaction in the bedroom. But the good-in-bed effect illustrates just how difficult it might be to resolve dissatisfaction, if each partner believes the other person is more responsible. You can read more about the good-in-bed effect at http://www.ncbi.nlm.nih.gov/pubmed/24003588.

The point of this essay was to wonder aloud, "where do our ideas come from?" In contradiction to what I've learned about quantitative social psychology, I did not conduct any research to answer this question.

Instead, I used my own experiences (an N of 1) as the basis for my analysis, and I published my results without attempting to replicate them. My suggestion, supported only be anecdotal data of a personal nature, is that my ideas have come from random processes and odd conjunctions of fate and coincidence. But like so many other published articles, my final statement is the clarion call for additional research: Readers, what are the root causes of your ideas?

References

Beggan, J. K. (1992). On the social nature of nonsocial perception: The mere ownership effect. *Journal of Personality and Social Psychology, 62*, 229-237.

Beggan, J. K., Vencill, J. A., & Garos, S. The good-in-bed effect: College students' tendency to see themselves as better than others as a sex partner. *Journal of Psychology, 147*, 415-434.

Beggan, J. K., Messick, D. M., & Allison, S. T. (1988). Social values and egocentric bias: Two tests of the might over morality hypothesis. *Journal of Personality and Social Psychology, 55*, 606-611.

Beggan, J. K., Gagné, P., & Allison, S. T. (2000). An analysis of stereotype refutation by an editorial voice in *Playboy*: The advisor hypothesis. *The Journal of Men's Studies, 9*, 1-21.

Zajonc, R. B. (2001). Mere exposure: A gateway to the subliminal. *Current Directions in Psychological Science, 10*, 224-228.

Sun–Planet Connections: The Problem of Boundaries of Knowledge

Janet Luhmann

Janet Luhmann is a Senior Research Fellow at the Space Sciences Laboratory, University of California, Berkeley. She works in areas of space physics where connections between disciplines-especially solar physics, planetary science, and astrophysics, is especially important to make progress.

Getting started

The trajectories each of us practicing some sort of scientific research travels is often highly circuitous and as individual as the practitioner. However, most of us have a memory of some event or experience or person that inspired us to choose a particular path. Such choices are affected by timing and circumstances, including prevailing conditions in the natural world and society, the state of civilization, and human knowledge. It was a definite advantage to grow up in the US during an era of expansion of the nation's investment in science and technology. The downsides of the cold war included air raid drills in school to prepare for potential atomic bomb attacks and scary TV videos of atomic bomb test destruction. But, at the same time, the media coverage of the bomb tests allowed the public to witness a highly visual demonstration of humans understanding nature at a sufficient enough level to harness incredible energy.

My father, a mechanical engineer by trade, took me out to our backyard at night to see Sputnik flying overhead-heralding the space age. The Apollo mission followed: a demonstration in its highest form of what human spirit, ambition, and dedication could achieve toward directing that knowledge to explore frontiers beyond the Earth. Witnessing events like these can easily plant seeds. It became not only possible, but desirable to pursue a career in space research. Graduate school was a chance to experience truly deep involvement in a scientific project, and watching a balloon carrying my cosmic ray experiment into the stratosphere over Hudson's Bay cemented my resolve to continue a life in space science.

The measurements from this experiment contained hints of a discovery, announced some months later by a more senior cadre of authors, that Jupiter was a source of some of these electrons – previously considered from outside the solar system. A lesson learned here was that new findings can be lurking anywhere if one has sufficient knowledge to recognize them!

Forging ahead

To move beyond the relatively sheltered environment of graduate school into a space research career is an adventure in itself. Space science is not a typical industry commodity, and in academic institutions where there is often a long list of physical science specialties that come in and out of demand in departments, space often falls in the gaps between physics, astronomy, and geophysical sciences departments. At least while the cold war continued, government investments in national and academic institution laboratories provided a home and resources for specific areas of work. NASA was a young agency generally viewed with awe and pride for its achievements. The space age gave rise to a host of Earth orbiters instrumented to characterize Earth's space environment, to survey resources, and inform national security. Space-borne telescopes for astronomical observations - free from the interfering atmosphere - became possible. There was much work to be done in both space mission design and in interpreting observations from space and from ground-based observatories.

The Space Sciences Lab at the Aerospace Corporation provided a chance to apply my knowledge to practical problems and to learn more about our local space and what influences it along the way. It was a great place to start because the Earth's space environment includes so much of what is relevant throughout the solar system and beyond. And its relative accessibility to measurements provides the ultimate 'ground truth' for the many less well-observed, distant, and impenetrable places in space of interest to us humans.

An intriguing new area of research that came together around this time was referred to as 'sun-weather' relationships. A group of researchers with a mix of backgrounds were investigating hints in data sets that the Sun was affecting Earth's near-surface atmosphere through mechanisms other than the standard solar photon flux. If this could be proven, it would change the way scientists viewed our relationship to our star. I began to think about ways one could investigate the physics of such coupling and wrote a failed proposal related to investigate atmospheric turbulence related to auroral activity, delaying my own further work on the subject. As potentially important as this question was, it remained (and still remains) at the edges of mainstream space research-in part due to its overabundance of correlations of sunspot number with everything from political events to investment success. Physical explanations of often transient or locally restricted atmospheric phenomena, based on the apparently small forces and subtle factors involved, remained elusive…as did broader support for this research area. Yet, this problem presented a perfect example of the requirement to think outside the box of one's own expertise –or run the risk of always seeking solutions to science questions in one's own bailiwick.

A next opportunity to expand horizons arose at UCLA in the form of data sent from a new mission orbiting Venus, our so-called twin Earth. To share the amazing privilege of being one of the first to look at new observations from a distant 'terrestrial' but very different world is a chance not to be missed. It is also worth mentioning how different it is to work with space observations that are not images. The 'wiggly lines' of time series and spectra, statistical studies, and dots on a graph can contain ground-breaking and paradigm-changing information but they need to be decoded to contribute to broader understanding. Learning to work with, extract information from, and share the results from these types of data is tantamount to learning a new language.

When the Pioneer Venus Orbiter got to Venus it not only imaged the surface beneath its obscuring clouds with radar, and the clouds themselves at various wavelengths of light, it also detected magnetic fields, the gases in the thin upper atmosphere, and a space environment so unlike Earth that the twin label seemed inaccurate for all save the planet's size and similar solar distance. While previous remote sensing and a flyby by an earlier interplanetary spacecraft had already given us expectations of such a situation, these could not prepare us for the full picture that the orbiter obtained in its nearly decade and a half long mission.

Here was a world where the atmosphere contained essentially all of the carbon dioxide present on Earth as limestone. The absence of liquid water on Venus, compared to Earth, was responsible. Its closer proximity to the Sun, together with its history of particular types of gases in the atmosphere, combined to make a 'runaway greenhouse' where the water was vaporized, split into its hydrogen and oxygen components, and either lost to space or reacted with the surface. Of special interest to me was how this happened, and in particular what occurred in Venus' atmosphere, now crushingly dense, intensely hot, dry and caustic, over time. It turns out that Venus is something like a comet in the way it interacts with the Sun and its outflowing upper atmosphere-the solar 'wind.' Venus has a comet-like tail of escaping atmospheric ions, including oxygen. This loss of oxygen could have played a key role in the fate of a possible early Venus ocean.

Meanwhile, the opportunity to conduct that delayed 'planet Earth' research project came my way. Using high latitude radars to observe downward propagation of auroral activity, we found that related disturbances in the upper atmosphere above Chatanika, near Fairbanks, penetrated down to at least 80 km altitudes when solar activity caused intense auroral storms. Whether these changes in our 'middle atmosphere' could provide a mechanism whereby solar activity also affects the lower atmosphere, perhaps by modifying conditions for the transmission of atmospheric waves coming from below, remains unresolved. The debate about 'Sun-weather' relationships continues to this day with brief flurries of excitement over new findings often giving way to frustrating lack of independent validation or reproduceable effects. Nevertheless, it is generally agreed that dynamical changes of the type we found occur – though the consequences remain unclear. New super computer models have since been developed that link the atmosphere from the ground to its boundary in space. Numerical experiments with these models may ultimately identify the mechanisms at work in this connection, and their overall consequences.

Moving to the Space Sciences Laboratory, at the University of California, Berkeley, completed the star-planet connection chain by opening the door more widely to the world of solar physics. The bigger picture of how planets and their stars interact was now in better focus.

Along the way has been the journey of a lifetime in space science. I have marveled at the Earth's aurora from outside Fairbanks Alaska, exploding in the sky over the ionosphere-probing radar used to measure its effects. I have collaborated with colleagues on Soviet missions destined for Mars -with special focus on seeking comparisons with what we saw at Venus. While these planetary neighbors of Earth are very different in appearance and conditions, they have in common a notable lack of water-and (perhaps by coincidence?) -both exhibit a comet-like removal of atmospheric ions related to their interactions with the Sun. I have seen the bright plumes and felt the roar from space launches of missions from KSC whose data I have helped mine, including the Cassini mission to Saturn and Titan, where exotic conditions contrast with (and are sometimes surprisingly similar to) those found on the terrestrial planets- though its Methane and Nitrogen dominated atmosphere is more like early Earth was thought to be than current Mars or Venus.

I have been part of an attempt to get a microphone to the surface of Mars for the first time, on the ill-fated Mars Polar Lander. The effort to realize this very human experiment continues in several groups now, never failing to gain new fans who inquire about the prospects for it every year. I have seen a total solar eclipse reveal the usually invisible solar corona in all its ethereal glory. (Related pictures can be found at the Exploratorium website: http://www.exploratorium.edu/eclipse/2006/viewer/viewer-how.html, highlighting their live webcast of which I was a small part (slide 32). A photo of the solar corona - http://www.exploratorium.edu/eclipse/2006/photo_4.html).

I have had the opportunity to lead a team providing science instrumentation for the STEREO twin-spacecraft mission that gives the first regular information about the invisible (to us) 'farside' of the Sun, and measurements of space environment conditions at other locations along the Earth's orbit. There we observed the equivalent of a 'Carrington class' event, which in 1859 was experienced on Earth and still holds the record of the strongest solar activity related magnetic storm in modern human history. Had this occurred about 10 days earlier when the activity would have been on the Earth-facing side of the Sun, it would have resulted in a remarkable natural experience for humankind and a test for our high-tech society. (The Carrington event was followed by auroras near the equator and spontaneous fires from large currents induced in long conducting cables for example. One can only speculate how the GPS system we rely on so heavily today would have behaved under similar circumstances, or how our modern power grids would have reacted.)

I have had the chance to participate in interpretation of observations from more recent European missions sent to our neighboring planets Venus and Mars, Venus Express and Mars Express. I have been involved in major projects coupling models of solar behavior to near-Earth space using large, integrated computer simulations - with the goal of both understanding, and someday predicting, the exchanges between the Sun, interplanetary space, and the atmosphere that determines 'space weather' effects we experience every day. Such endeavors are similar to the breakthrough that occurred when Earth's surface features and oceans and clouds were added into weather models –with major consequences. I am part of a team examining results from a new mission to Mars called MAVEN, designed to investigate what role the Sun has played in the loss of that planet's early atmosphere, and with it, the once much warmer, wetter environment there. The new measurements are in the process of being gathered, analyzed and interpreted as this article is written- and will contribute to an unprecedented new picture of the Mars upper atmosphere and its interactions with the Sun above and the ground below. The interested reader can follow the progress of discovery during the MAVEN mission at http://lasp.colorado.edu/home/maven/.

The Science Feast that is the Solar System

The early reconnaissance of the solar system carried out with the Mariner, Pioneer and Voyager missions, and their non US counterparts-many from Soviet probes to Venus and Mars –hinted at the veritable smorgasbord of strange worlds, moons, conditions and phenomena for humans to ponder. These explorations brought home appreciation of the specialness of the Earth among planets. At the same time, we learned much more about the star that gave birth to our planetary system and whose character sealed our fate, and continues to do so. While most humans only experience the Sun as a predictable, warmth-providing bright disk in the sky that rises and sets daily, early observations of the solar corona at the rare times of local total solar eclipse, and the auroras observed mainly in the polar regions suggested that much broader interactions exist. In particular, the invisible solar wind and the magnetic field it carries transfers energy and momentum into Earth's space environment and upper atmosphere whose short and long term effects are still being explored.

Observations in Earth orbit started providing awareness of our home in space on a regular basis. The internet has now given virtually every person routine access to these wonders formerly accessible only to scientific and technical specialists. One can easily find a daily 'space weather' report showing pictures of today's corona seen by spacecraft cameras (for example, see spaceweather.com and solarmonitor.org and their related links), or information on the next lunar or solar eclipse, interesting planet arrangements, meteor showers, or the occasional comet in the local night sky. There are even 'citizen science' sites where anyone with a good WiFi connection and interest can use state-of-the-art space science data to make real contributions to its analysis and interpretation… in a form of both crowd-sourcing and inspiration-sharing.

The related changes that have occurred in science in recent decades have been equally transforming. Those in a research career face a huge menu of problems and projects to choose from, though as a practical matter, most find their choices directed into areas that peers and sponsoring agencies wish to support, arising from priorities set by preceding generations. And herein lies the rub. Sub-disciplines develop that can limit broader progress in knowledge and paradigm-changing perspectives. There is so much detailed work being done and information disseminated - compared to the beginnings of the space age - that synthesizing and absorbing it is beyond a single human's capability. Our challenge is how to move forward effectively on big picture, paradigm changing levels of science within the framework of this abundance of riches. Big picture progress keeps us on track to where we need to go next, toward better knowing our place in the cosmos. After all, that is why we explore space.

A favorite subject of mine in this regard is our failure to give high enough priority to deducing the history of the Sun. One of the most essential missing pieces in reconstructing the history of our solar system is the lack of knowledge of solar evolution and its many effects. From its formation from an interstellar cloud to the present, the Sun has played a host of transformational roles, most of which we can only imagine from observations of Sun-like stars at earlier stages of their evolution. After solar radiation and the early solar wind cleared out the gas and dust debris left over at the end of planet formation, the Sun must have influenced, and often dominated, the states of the planets and their atmospheres. This fundamental problem of space science requires a combination of expertise in stellar evolution, planet evolution, and the interaction of the two, together with new thinking on records buried in solar system small bodies and the Moon. We cannot say we understand the terrestrial bodies of exoplanetary systems until we understand our own. Perhaps the new generation will embrace this and other identified yet unaddressed big challenges, with many of their own discoveries along the way.

A Leopard's Breakfast Interrupted: Further Explorations of a Congolese Wilderness

Thurston Cleveland Hicks

Thurston Cleveland Hicks is a primatologist at the Max Planck Institute of Evolutionary Anthropology in Leipzig, Germany. He is currently teaching classes on the relationship between humans and the other great apes in the faculty of Artes Liberales at the University of Warsaw in Poland.

Introduction

I was born in Raleigh, North Carolina in 1972. During my first five years of life, my family lived in the big city, but then we moved out into the country, which allowed my brother, sister, and me the opportunity to run free and explore the forests, streams, and clearings surrounding our neighbourhood. My mother and father educated me in the virtues of conservation. On one occasion my brother Walker and I discovered a vulture nest in the crumbling recesses of an old abandoned barn. We had seen the same vultures striding by the barn a short time earlier, regal feathered dinosaurs scanning us with keen and imperious eyes. When we proudly brought one of the eggs back home as a prize, my mom scolded us and told us to return it to its nest as quickly as possible, lest it break or the mother vulture be forced to abandon it due to its human smell. I never forgot my mother's words, and the valuable lesson I learned that it is not necessary to own something or to alter it in order to appreciate its existence.

After a brief stint working as a volunteer field palaeontologist during my undergraduate years, I decided to study living beings instead. I received my undergraduate degree in Anthropology from the University of North Carolina at Chapel Hill, then spent the next 10 years studying everything from ring-tailed lemurs at the Duke Primate Center to rhesus monkeys on Cayo Santiago, Puerto Rico, and finally western lowland gorillas for 2 years at the Mondika research site in the Central African Republic. I acquired my Master's Degree studying the chimpanzees and gorillas of the Ngotto Forest at Central Washington University under Professor Roger Fouts. Beginning in 2004, I devoted the next 10 years of my life to studying the chimpanzees of Bili, Democratic Republic of the Congo, first at the University of Amsterdam where I defended my Ph.D. thesis under Professor Steph Menken, and then at the Max Planck Institute of Evolutionary Anthropology in Leipzig, Germany, in the Department of Primatology lead by Professor Christophe Boesch where I am currently based.

August 5, 2012. Bili, Northern Democratic Republic of the Congo, Central Africa.

We struck out from our temporary camp in Mister Ginigbya's abandoned manioc field just after dawn, eager to reach Camp Gangu and begin setting out our camera traps. During the night a tropical storm had struck and the forest was drenched. A light drizzle would continue to fall for most of the day, soaking into everything from my field notebook to my socks and keeping them dank and soggy. Nevertheless, our team was infected with a sense of excitement about the discoveries ahead. Just the day before, we had left the high grasses of the savannah behind us and crossed into the eastern edge of the Gangu Forest. I was eager to see if it remained the remote wildlife bastion it had been when I had last been here in early 2007. Encouragingly, we had seen the tracks and dung of multiple elephants moving south across the savannah behind us, and I had high hopes that the forest might have remained somehow isolated from the poachers and miners responsible for so much destruction elsewhere.

I had been coming to this remote forest since 2005, drawn across the savannahs and swamps by the astonishing tales of our local Zande field assistants: they told us of a vast primeval wilderness criss-crossed by elephant 'superhighways,' where leopards would still mistake people for red river hogs and chimpanzees would approach visitors on the ground with a spirit of innocent curiosity. In these days when our global society seems to worship frenetic development, road-building, and 'resource-extraction' above anything else, such places have become vanishingly rare, and on our first trek in 2005, when this hidden world had first unfolded around me, I had hardly been able to believe what I was seeing. When first gazing up into a tree-full of Gangu chimpanzees, I was stunned that they did not shriek and leap to the ground in terror, which has now become the normal behaviour for their kind throughout Africa. Such fear of humans is not surprising given that chimpanzees are being ever more relentlessly persecuted, hunted for their bushmeat as well as for their babies to be sold on the open market as pets. Many of these Gangu chimpanzees, unlike those I had encountered near the roads, would regard me with mildly curious, almost blasé expressions, much as they might cast down towards a passing baboon or warthog. Some of them would move out into the open to scrutinize us, but they would continue chewing languidly on their minty green *Parinari* fruits. Some of the youngsters would actually move towards us along the branch. Once an adult male had approached us on the ground to within a few meters, where he sat glaring at us with an intensely furrowed brow. The experiences these chimpanzees had had with humans in the recent past must have been limited mostly to watching from the trees as small Zande families padded by barefoot, armed with nothing much more potent than a crossbow with a few arrows, on their way to the nearest dry streambed to dig for aestivating catfish. The naïve reaction of these chimpanzees to *Homo sapiens*, who in other contexts is their most fearsome and relentless enemy, is a bit shocking to see. It also presents us with an excellent opportunity for research and conservation.

My passion for the past 20 years has been the study of chimpanzee behavioural variation across Africa, which has been described by primatologists as a kind of non-human 'culture.' We know from fossils and genes that around 5-7 million years ago, one group of chimpanzee-like creatures split off from its tree-dwelling cousins and set off down the evolutionary pathway that would eventually lead to us. What were the factors that pushed our ancestors to separate from the chimpanzee lineage? Many theories have been put forth, including changes in climate, habitat, social structure, and diet, but regardless of the context in which this happened, it seems quite likely that the initial differences between the two otherwise similar ape species might have been behavioural, or even cultural - perhaps a fondness for this over that kind of prey, a knack for cooking meat above the crackling embers of natural savannah fires, or even a new kind of (bipedal?) display useful in attracting the opposite sex. For this reason, it is crucial to understand the extent of behavioural variation found today in modern chimpanzees and other great apes.

It just so happens that chimpanzees, which are, together with bonobos, our closest cousins, have the most extensive and flexible repertoire of tools and food extraction techniques of any non-human species to have ever been studied. Whenever a new population of chimpanzees has been encountered, they have always revealed new patterns of behaviour, from the use of stone hammers to crack open nuts in Ivory Coast to the fabrication of pointy-ended 'spears' used to skewer bush babies in Senegal. In addition, chimpanzees everywhere are eager and resourceful hunters of a variety of prey, from duikers to monkeys, with up to 5% of their diet being composed of meat. In the Taï Forest they have been observed pursuing red colobus monkeys through the canopy in coordinated teams. Such discoveries have blurred the lines which once so solidly delineated humans from the rest of the animal kingdom. It used to be assumed, for example, that only humans manufactured and used tools, and that humans were the only primates to practice organized hunting. These claims for human exceptionalism are no longer tenable. The diversity of chimpanzee material culture resembles that possessed by our own ancestors. For a primatologist entering an unexplored forest full of chimpanzees, the learning curve is sky-high and the sense of imminent discovery is palpable.

One of the great frustrations for a field primatologist such as myself is the difficulty of observing our study species without affecting the animals' behaviour. When your subjects cease their activities upon spotting you, begin defecating in panic, or simply vanish silently into the undergrowth, you feel not only guilty for having intruded but also a sense of despair that you will never achieve more than a surface understanding of the rich but hidden lives of these apes. For that reason, researchers spend years 'habituating' great apes to human presence, at great cost and effort. This would not be an option for us at Gangu given the political instability of the Democratic Republic of Congo - it is ethically questionable to familiarize chimpanzees with humans when we cannot assure their safety from poachers. Up to this point, our studies of the chimpanzees of the Bili Forest had been largely limited to the artefacts they left behind: ground nests, stick tools used to dip for ants, and smashed termite mounds and snails. For this reason the 'naïve' behaviour of the Gangu chimpanzees was a research boon. It was not that the chimpanzees always continued on with their behaviours unaffected when they saw us. It was just that they were much less wary. Even on occasions when they had seen us and moved off, they would stay in the area and continue more or less the same activity - we would often find them in the same place the next day.

It was this relative insouciance that had allowed me to sneak up on a group in 2005 and film them dipping into an ant hole with stick tools. On the same day, I listened to them at close range as they squabbled over the carcass of a tree pangolin. It was deep in the Gangu Forest that seasoned tracker Ligada would later be able to approach and witness a chimpanzee snacking on the carcass of a leopard.

This unparalleled opportunity to observe fearless chimpanzees was one reason why, armed with camera traps and GPSs, we had returned to this dream-world to fine-tune our knowledge of the behavioural traditions of the Bili apes. But we were here as well for an even more important reason. I had begun my career as an academic, a student of human evolution, who had flown to Africa searching for a pristine, remote wilderness. I had been seeking a forest world still ruled by great apes and other wild things, unaffected by the digging, cutting, razing, and burning of nature carried out daily by millions of busy human hands. The two field sites where I first worked, the Ndoki Forest in Northern Republic of Congo and the Ngotto Forest in Central African Republic, were spectacularly rich habitats for mammalian mega-fauna, but each had been becoming rapidly encircled and penetrated by a metastasizing network of roads, logging and mining camps. Elephants had already been eliminated from Ngotto, and a logging road had been installed within 10 km of the Mondika Base Camp at Ndoki (as I write this, the conservation world is holding its breath as ivory poachers begin a massacre of forest elephants just to the north of Mondika at Dzangha Bai). In the end, we had been forced to abandon our efforts to habituate the Ngotto chimpanzees when I encountered a logging survey team emerging from the forest with their backpacks stuffed full of bushmeat. Was there anywhere left where the 'genie' of development had not yet been unleashed from the proverbial bottle?

When I arrived at Bili in 2004, having been invited by the NGO, The Wasmoeth Wildlife Foundation, and conservationist, Karl Ammann, I thought that I had finally found such a place. The human population density at Bili was extremely low, and vast tracts of savannah and forest remained completely unpeopled. In their place roamed a fascinating mix of savannah and forest species: elephants, chimpanzees, giant forest hogs, and golden cats cheek by jowl with warthogs, hyenas, buffalo, baboons, and lions. The local Zande people certainly impacted their environment, burning the savannahs once per year, digging up fish from dry streams and bow-hunting along the roadsides, as well as practicing slash-and-burn agriculture, but this disturbance was mostly limited to areas adjacent to Bili's one major north-south road. At a distance of about 20 km west from that road, which was the eastern boundary of the Gangu Forest, we began to encounter pile after pile of elephant dung; not coincidentally, there our encounter rate of human tracks and trails dropped to nothing.

Over the next two years, we explored this forest and eventually established a base camp in its core. I was able to document an intriguing chimpanzee culture new to science. The Bili chimpanzees, like gorillas but unlike most other populations of chimpanzees, have a penchant for spending the night in ground nests – 19% of their nests in the Gangu Forest were on the ground. This was especially surprising given the high density of predators in the forest, together with other dangerous big mammals such as elephants and buffalo. The nests were unsually large elaborate affairs, made from multiple branches ripped off or bent in from nearby saplings and carefully woven into a springy bowl-shaped structure. Another quirk of the Bili population was their behavior of smashing open *Cubitermes* and *Thoracotermes* termite mounds against roots and buttresses, presumably to eat the wriggling grey termites inside. Interestingly enough, they ignored the omnipresent red dome-shaped

Macrotermes termite mounds scattered across the landscape. In a number of other populations, chimpanzees avidly construct 'fishing wands' with which to extract fat and tasty *Macrotermes* soldiers for consumption. Could this difference be ascribed to culturally different tastes or tool-making traditions? The Bili chimpanzees did not limit themselves to pounding open termite mounds, either. We also found the smashed remains of hard-shelled fruits, African giant snails, and even tortoises at their 'workshops' scattered about the forest. Finally, these chimpanzees showed an unusual profile of preferred prey species: a tree pangolin and even a leopard were observed being consumed!

Some fortuitous constellation of factors had somehow prevented humans from settling in and spoiling this enchanting place. Upon my return to the University of Amsterdam, where I was writing up my doctoral thesis, I made as much noise as I could to alert the world to the existence of Gangu. I had high hopes that we could draw in a team of wildlife guards to protect this vulnerable natural jewel, and then invite in a team of scientists from all over the world to study the fauna and flora out of Camp Gangu. But we were not quick enough.

In July 2007 I received the heart-breaking news from Wasmoeth Wildlife staff that, within a matter of weeks, several thousand gold miners had poured into the Bili-Uéré Protected Area Complex and set up shop in two now bustling gold mines about 60 km northeast of the Gangu Forest. Our long-term plans for the area were immediately scuttled. In a harrowing night-time flight, Wasmoeth Wildlife staff escaped from the now-hostile town of Bili in the project Unimog. Bili Town descended into chaos. I would later read a police report describing the burning of the central Bili government center, and we heard first-hand accounts of looting, murders, and riots. Elephant poachers took advantage of the chaos to ply their destructive trade in the Gangu Forest. Unable to return to Bili, I spent the next year and a half conducting wildlife surveys about 200 km south, on the other side of the mighty Uele River. Intriguingly, despite living in a more continuous forest cover than to the north with some differences in vegetation and the cast of animal characters (there, unlike at Bili, okapis and red colobus were present, but lions and hyenas were not), the South Uele chimpanzees seemed to share the same suite of behaviours as those at Bili: frequent ground-nesting, the smashing of termite mounds and snails, ant-dipping, and a lack of termite-fishing. Although this discovery added a whole new element to my thesis, I also stumbled upon a developing conservation crisis: during my time there, we saw 42 chimpanzee orphans and 34 carcasses for sale in markets, towns, and along the roadsides. In addition we saw okapi skins being used to make deck chairs and church drums, and elephant ivory and meat on open display. At the same time as traders were pedaling bike-fulls of domestic goats and chickens into the area, other merchants were busily biking out the indigenous fauna in crates and sacks. The sinister fingers of the commercial bushmeat trade were working their way into these most remote forests, facilitated by and associated with artisanal mining enterprises.

Tragically, it was probably too late to do much for the embattled fauna of South Uele, but at Bili to the north, we still had a fighting chance. In late 2010, The Lukuru Wildlife Research Foundation (LWRF) sent a survey team accompanied by guards to Bili. A bit over one year later, having acquired a substantial grant from the US Fish and Wildlife Service, I returned to Bili, leading a collaborative mission between a number of institutions: LWRF, The Max Planck Institute for Evolutionary Anthropology (MPI-EVA), African Wildlife Foundation (AWF), Lucie Burgers Stichting (LBS), and the Institute Congolais pour la Conservation de la Nature (ICCN).

That is how, in August 2012 at the height of the rainy season, I found myself sleeping in an abandoned manioc field at the edge of Gangu Forest, accompanied by six Kalashnikov-wielding conservation guards, four Congolese researchers from LWRF, MPI camp leader, Karsten Dierks, from Germany, and a small army of about 50 local trackers, cooks, and porters.

The field in which we had passed the night had still been active when the Lukuru field team first encountered it during their 2011 survey. Prior to entering the forest this year, we had encountered its owner, Mister Ginigbya, on a number of occasions while meeting with Bili officials. He was a lanky and laconic older gentleman with a proclivity for wearing multi-coloured shirts and a rakish zebra-striped cowboy hat. He had been present at most of our meetings with Chief Zelesi-Etienne, the head of the Gwamonge Collectivity in which much of the Gangu Forest is located. Zelesi had introduced him as the man who was working to re-establish the defunct village of Kalé in the heart of the wildlife reserve.

When commercial agriculture was implemented into the area about a hundred years ago by cotton-pickers working for the Belgians, Bili-Uéré had no protected status. According to our assistant, Chief Mbolibie Cyprien, who was a child at Bili prior to Congolese independence, at the time of his youth around 1000 Congolese settlers lived at the eastern edge of Gangu Forest between the Bo and Wolu Streams. A smaller colony of about 500 workers farmed cotton about 15 km east of that at the Lumbi River. The Belgian overseer who lived in Bili only occasionally visited the remote forest outpost. According to stories Mbolibie heard as a child, this overseer treated his workers harshly. He insisted on being carried along the Gitambo Road to Kalé on a sedan chair, and if one of his porters had the misfortune to stumble over a termite mound hidden in the high savannah grass, a hippo-hide whip would immediately strike down against the poor man's back. Today all that is left of the Belgian presence is a tangled regenerating forest, some oil palms and lemon trees, and a series of ancient road markers left along the overgrown Gitambo Road, which reaches its end at this site.

According to Mbolibie, the Groupement of Kalé itself was disbanded in 1964, during the time of the Simba rebellion. People remained in the region harvesting cotton until 1975, when the Belgian project came to an end, leading to a final mass movement of people to settlements alongside the road. Since then, the forest has been uninhabited and only occasionally visited by small numbers of Azande on hunting and fishing treks. As we documented during our 2005 transects, the Gangu Forest west of the Bo Stream showed no signs of any past human settlement over at least the past century; it is mostly primary forest interspersed with small patches of savannah, lacking any oil palms or regenerating vegetation.

This long period of seclusion ended a few years ago. With no active conservation presence in the area and a new wave of gold miners and ivory merchants operating out of Bili, a movement was undertaken to reclaim the forest, supported by Chief Zelesi and led by Mr. Ginigbya, a direct descendent of the Kalé family. Two fields were cut into the heart of Gangu, to grow manioc and other crops. Due to the recurring threat of insecurity in the region, however, this movement was stillborn. Around February 2012, five heavily-armed men wearing military uniforms originating from the neighbouring Central African Republic had invaded Gangu, periodically emerging from the forest to ransack fields and villages for food and terrorize Bili-ites at gunpoint. They were commonly referred to by the people with whom we talked as the 'LRA' (the Lord's Resistance Army of Joseph Kony), but the affiliation of these mysterious invaders was quite unclear, along with where they might have acquired their weapons. All that was known about them was that they had facial features typical of the northern Sahel peoples and they spoke little to no Lingala (the trade language used across north-central DRC). Following a number of violent clashes with Congolese troops (FARDC) and local militia, these bandits finally fled the Bili area; we were told that two of their number were killed. While in the forest, they had attacked Ginigbya's field, briefly holding his son hostage and causing his family to flee and abandon their recently-planted crops to the local fauna…for the time being, anyway.

The manioc field in which we had camped that August evening was rapidly going fallow; elephant feet had trampled the surrounding vegetation and most of the manioc had already been devoured, which was a bit of a shame, as we could have used some variety in our diet after several days of mostly beans. We found four huts already crumbling beneath the weight of thick lianas and fallen trees. Farming had clearly not been the only activity here: inside one of the huts amongst old moldy dishes and rotting clothes was a pile of eleven shotgun cartridges, and we found six vines looped around the rafters of another hut which had been fashioned into crude snares. This was a disturbing development: on our 2005 transects we had seen not a single sign of hunting or snaring in the Gangu Forest. At least we found no pans and shovels for gold mining here.

Other than this, and a bout of nausea which floored me for a few hours, our stay at the camp had been uneventful, and other than clearing a patch of field to make space for our tents, we would leave it as we found it. This would not stop rumours from later being spread around Bili by Mr.

Ginigbya that our ICCN guards had maliciously burned his huts and ravaged his fields. These baseless rumors would be accompanied by demands for financial compensation. The destruction of his fields had, however, already been accomplished months before our visit by an industrious team of red river hogs and other herbivores.

Setting off through the damp forest, we worked our way southeast along an old fisherman's trail. Red tailed guenons and agile mangabeys chirruped and squawked in the trees above us as we passed. We encountered a disturbing number of trees marked with the graffiti of known elephant poachers: Shimita, Dieu, Zapones, and Romain. Like the manioc field, this was something we had never before seen in the Gangu or even the Camp Louis Forests during our extensive surveys five years earlier. In the days ahead, often when we would find the passage of an elephant ploughing through the forest, or cross-one of their well-worn trails, we would also see machete cut marks right behind them, undoubtedly left by poachers tracking their quarry. How had the density of mammals at Gangu, which had been most impressive during our earlier field surveys, been affected by this increase in human activity? This was one of the main questions we were trying to answer on the current mission. I had for the last few years been tormented by nightmares that the forest would have been converted to cookie-cutter style American suburban neighborhoods in our absence.

We slowly ventured along the slippery path. Ephrem Mpaka, one of the four Lukuru Foundation researchers accompanying us, periodically vanished from sight up ahead. I would wipe the fog off my glasses and try to figure out where he had gotten to; a minute or so later he would pop grinning out of the greenery ahead onto the trail, then ebulliently describe an animal path he had just followed to a decaying pile of buffalo or elephant dung, duly photographed. We diligently documented all of the animal evidence we found, taking photographs and recording GPS waypoints. Several days prior, in the savannahs near old Camp Louis, we had encountered the recently butchered and smoked remains of a warthog and a forest buffalo, and we had been worried that we might find an empty forest ahead. Happily, though, the signs were still plentiful: elephant, buffalo, chimpanzee, leopard, even (in the days to come) giant forest hog and hippopotamus.

Around noon, as we approached the rain-swollen Bo River, we were obligated to pick our way across a small swampy creek, the Dziliwo. The mud sucked at our boots, and we ducked repeatedly to avoid dangling thorns and cruelly barbed spines. We arrived at an area where the swamp mud had been churned up by what appeared to be some violent skirmish. Herbs had been uprooted and flung about, and gaping red slash marks defaced the base of a tree. Something momentous had happened here at the Dziliwo, apparently in the predawn hours following the heavy rains of the previous night. Ephrem darted ahead to point out the welter of deep gouges left by red river hog hooves in the soft mud. A large number of these tusked forest pigs (scientific name: *Potamochoerus porcus*), with their characteristic wizard-like visages and dangling white eartufts, appeared to have engaged some foe in a pitched battle, racing back and forth and trampling the vegetation. The identity of that foe was quickly revealed. Lukuru researcher, Bebe Bofenda, pointed out the large rounded prints of a leopard (*Panthera pardus*) squashed into the mud inches away from one of the hog prints. Fresh dung of both species was also present at the scene.

Bebe's gaze drifted upwards into the trees, and I saw his face freeze in astonishment. '*Monsieur Cleve, regardez-la*!' he whispered. My eyes followed the line from his finger up into the dripping wet canopy and I was stunned to observe the headless carcass of a red river hog sow, suspended 40 feet (approx. 20 m) above us in the crook of the tree where the leopard had stashed her. Her pink legs were splayed out stiff in an almost supplicating gesture. Fragments of the sow's jaw were scattered on the ground at our feet, beside a coil of leopard dung –the big cat had chosen to eat the head first. Possibly even as it was relishing its meal we had arrived and startled it off. It had left the rest for a later meal. The pig's jaw had the soft bones and dentition of a young animal.

Investigating the site further, we found a large number of fresh hog pellets at the base of a nearby stilt-rooted *vwula vwula* (*Uapaca*) tree, along with more trampled vegetation. Etched into the trunk of this tree we saw vivid red gashes where, according to the trackers, the hogs had vainly tried to attack the leopard prowling above their heads. Bebe pointed out another set of gashes higher up the tree, three in parallel, made by the leopard's claws. I could picture it in my mind's eye: the big cat snarling down at the circle of enraged hogs from its perch just out of reach of their razor-sharp tusks. From there, if the trackers' interpretation of the evidence was correct, it must have hauled its still-quivering booty into the canopy of the neighboring tree and settled down to its feast, ignoring the cacophony of snorts, squeals, and roars issuing from its sworn enemies below.

It was difficult if not impossible to determine from the evidence before us the precise sequence of events which occurred during the attack. Had the leopard seized the sow first, carrying her up into the tree before her groupmates arrived in response to her frantic squealing? Or had the cat pounced on its prey in the middle of her group, killing her before the other hogs could react and then escaping from the ensuing chaos into the canopy above, only returning to the forest floor to retrieve the carcass when the herd had left? We excitedly discussed the evidence before us and struggled to recreate the conflict we had possibly just missed. The trackers insisted that the deep gouges at the base of the tree had been made by the hogs, and we had seen similar traces before.

We wrapped up our snapping of photos and scribbling of data and moved along down the trail. This scene to me served as a dramatic confirmation that Gangu had survived to this day as a functioning ecosystem, with the age-old dance between predator and prey continuing to play itself out morning and night. As icing on the cake, only a few meters south Ephrem pointed out a fresh elephant print. My fears that during our 5 year absence the Gangu had been converted into a gigantic manioc plantation receded into the mist as we walked along. I was in great spirits - this was after all only our first full day in the Gangu Forest. What else would we observe after we settled into our base camp in the 'Couer de Gangu'?

As we moved away from the Dziliwo, I was startled as a camouflaged shape emerged from the shadows behind a tree just at my heel, and I briefly thought that perhaps it was the leopard, returning to reclaim its kill site from these impertinent intruders. No, it was Feruzi

Yenga, one of our six Ecogaurds. They had fanned out across the perimeter to guard us against precisely that possibility. I complimented Feruzi on his quiet forest stealth, and we headed onwards towards Camp Gangu.

Finally, after having crossed the Bo River and hiking through several more kilometers of elephant and monkey-rich forest, we arrived in the late afternoon at the New Camp Gangu, which had been established by the Lukuru team in 2011. The sun had finally emerged and we had dried out a bit, but we got wet all over again when we crossed the waist-deep Gangu River. A few months later at this very spot, Karsten and team would find fresh hippopotamus sign! Hippos are not found in this forest during the dry season, but the flooding of the Gangu allows them to expand their range, even into the savannas to the east.

This camp was located about two kilometers southeast of my old Camp Gangu established back in 2006. In 2011, Henri and team had found an active manioc field here. The owners had not been present, and the Lukuru team had fashioned it into their own temporary base camp. It had been necessary to relocate here from my original camp because at the height of the dry season, it was the only place which would still have accessible water. We had learned at Baday Village that the field, now abandoned, had been planted about two years ago by a now-deceased fisherman and his family to serve as a supplementary source of food during their periodic fishing trips to Gangu.

A lot of work must have gone into clearing the field, requiring the felling, with simple axes, of dozens of centuries-old trees. All of this had come to naught, however, with the death of the family patriarch due to an illness. A few months later, our former tracker Likango and another fisherman had been attacked while camping at this site, by the same five mysterious Central Africans who had ravaged Mr. Ginigbya's field to the northeast. They had struck Likango over the head with one of their weapons and stolen all of the fishermen's food and equipment. No one had camped here since - the shacks were rotting and carpeted with lush mats of riverside herbs. Inside one of the decaying huts, we found a few dishes and a single red shotgun cartridge. Luckily for us, the manioc plants the original owners had planted were still thriving and in the weeks ahead the leaves and roots would supplement our tedious diet of beans, day in and day out, beans.

As would be the case with Mr. Ginigbya, the deceased man's family would later claim that our team had 'stolen' their abandoned field and would demand recompense. The awkward question must, however, be asked as to whether the local people have a right to cut open large fields inside a forest classified by the Congolese government as a protected 'Domaine de Chasse,' especially when there is clear evidence that hunting and snaring are a part of that occupation. Of course, the traditional chiefs and the Congolese government have different opinions on this, and even within the Congolese government the position is not unanimous. We have good reason to believe that some members of the FARDC, the Congolese military, are sending heavy-duty weapons into this and other forests to poach elephants. It may turn out that our presence here could even help make the forests more safe for small-scale fisherfolk and indeed for the Bili population in general, who do not want armed strangers using this forest as base from which to attack them. What a complicated web of conflicting and sometimes intersecting interests we will be obliged to work out!

Our porters got to work clearing the undergrowth for the installation of our tents and fabricating ingenious vine-and-stick stools and benches. Later, as the evening descended, I gazed out across the languid tea-colored waters of the Gangu River and retraced in my mind the long path that lay behind us between the provincial capital of Kisangani and the Gangu Forest. I imagined that just beyond us in the forest somewhere a leopard was reposing on a branch digesting its satisfying breakfast of hoghead, after having been interrupted by our pass-through; an elephant might be nearby, sampling tender new leaves wrenched from a *banga* branch with its dextrous trunk; and a drowsy chimpanzee was beginning to weave branches and herbs around itself into a springy ground nest in which to pass the night.

Chimpanzees, of course, were always the focal point of our study, and even now we begin to sketch out our plans to explore their nesting sites and gather and measure their ant dip tools. In the days ahead, Karsten would begin to put out his camera traps. Ephrem, Gilbert, and I, accompanied by guards, would fan out on a series of recces (exploratory reconnaissance walks) on both sides of the Gangu, and eventually repeat a key link of my 2005 transects to see if the chimpanzee population had remained stable. We hoped that the presence of the ICCN guards would provide the heavily-persecuted elephants a safe haven for a year and give them a little breathing room. Henri and Bebe planned to depart in a few days to survey the forests and savannas east of the Baday Bili road, called Dume. After an absence of 5 years, I was more than ready to fill up some field notebooks with data and stories.

As we approached through the forest towards the main Bili road, we were faced with the most formidable and unforgiving tangle of vegetation we had yet had the misfortune to hack and clamber our way through. Unlike the virgin forest to the west, this dense and unforgiving wall of vines and spikes had grown up following a century or more of repeated clearings for slash-and-burn agriculture. For the past month, our team had been repeating lines of the 2005 survey, our goal being to compare encounter rates with chimpanzee nests, elephant dung and other signs of big mammals with what we had seen before. I was quite excited, as we had just confirmed that the Bili chimpanzees use sticks to dig up underground bees nests and presumably eat the honey and larvae. Even more exciting, we had counted more chimpanzee nests than we had in 2005, and seen fearless chimpanzees who had remained peering at us from the treetops for more than half an hour. Finally, our team had managed to take some stunning photographs of a brooding adult male Gangu chimpanzee far to the west.

Alas, the elephants had fared much worse than the apes during our absence. Dung encounter rates were down by more than half. Worse, we had found three poached carcasses of elephants being gnawed into dust by termites. Two of these skeletons, lying in the damp leaves at a 2 year old abandoned poachers' camp, appeared to belong to an adult and a juvenile … probably a mother and baby butchered together for their ivory. According to our local trackers, a large number of poachers had swarmed into the forest just following the departure of Wasmoeth's team in 2007. In just a few short years, one of The DRC's last remaining elephant strongholds had been irreparably impoverished.

We emerged from the forest exhausted. My legs were etched with dozens of fresh scars and several of us were recovering from dehabilitating attacks of malaria and flu. I was left with a hacking cough and the lingering weakness left by a receding wave of malaria.Walking transect lines during the rainy season was perhaps the most difficult job I have ever done. There was good news, however: other than the elephants, the fauna of the Gangu Forest appeared to have remained relatively undisturbed since my absence. Karsten would soon send out some stunning camera trap videos: of elephants, hyenas (in the forest!), leopards, agile mangabeys, golden cats, and even chimpanzees using tools. Gangu was still a forest well-worth protecting.

Little did we know, however, that the months ahead would see a rapid deterioration of the situation. The aftershocks of the political chaos in the neighboring Central African Republic would send thousands of refugees pouring into Bili, and during the dry season poachers would once again follow the dwindling herds of elephants as they retreated towards their primary water source, the Gangu River. Our team would be forced to evacuate Camp Gangu due to threats from poachers, and upon Karsten's return a few weeks later, he would find the camp burned to the ground, including most of the chimpanzees' ant and honey dip tools which our team had painstakingly gathered. According to Chief Zelesi, this was the work of the elephant poacher, Romain, who had been arrested by our team of ICCN guards one year before. Other sources, however, would tell us that it was the chief himself who had ordered the attack on our camp, angry about the confiscation of one of his guns by a guard in the forest. We may never know the true story. Back at the Max Planck Institute in Leipzig, Germany, my main priority now is to get out the message that Gangu is still there, and still rich and teeming with beautiful fauna. We can no longer, however, take it for granted that it will still be there tomorrow, if we cannot find a way to protect it.

Here is my final mission report to US Fish and Wildlife - http://tinyurl.com/bilireport

Thurston Cleveland Hicks

Two scientific papers resulting from our study:

http://www.sciencedirect.com/science/article/pii/S0006320714000044
http://dare.uva.nl/record/1/332289

Acknowledgements

This mission was made possible by the generous support of the Max Planck institute for Evolutionary Anthropology, The Lukuru Wildlife Research Foundation, The US Fish and Wildlife Service, l' Institut Congolais pour la Conservation de la Nature, The Lucie Burgers Foundation, and The African Wildife Foundation. Special thanks to John and Terese Hart, Henri Silegowa, Ephrem Mpaka, Gilbert Paluku, Bebe Bofenda, Christophe Boesch, Hjalmar Kuehl, Paulin Tshikaya, Steph Menken, Jan Sevink, Peter Roessingh, Jan van Hooff, Roger Mundry, Sandra Tranquilli, Judy Song, Cosma Wilungula, Jef Dupain, Andrew Fowler, Laura and Adam Darby Singh, Ligada Faustin, Mbolibie Cyprian, Likambo, Seba Koya, Kisangola Polycarpe, Karsten Dierks, Claudia Nebel, Mimi Arandjelovic, Mizuki Murai, Dirck Byler, Richard Ruggiero, Jeroen Swinkels, Hans Wasmoeth, and our brave team of ICCN guards: Brigadier Ekunda, Brigadier Feruzi OPJ, Feruzi Yenga, Gazuwa, Bernard, and Tyson . Kitty Hicks, Walker Hicks, and Judy Song provided additional valuable comments on the text. Finally, without my family's support I would never have been able to go out and explore the world: many thanks to Susie, Jamie,Will and Luke Sneeringer, Katherine, Walker, Savannah, Isaiah and Solomon Hicks, and my parents Thurston and Kitty.

Lukuru Foundation youtube site with camera trap films of the animals of Gangu - http://tinyurl.com/lukuruvideos

See in particular:

Leopard: http://tinyurl.com/lukuruleopard
Hyena: http://tinyurl.com/lukuruhyena
Gangu chimpanzee tool use: http://tinyurl.com/chimptool
Gangu males: http://tinyurl.com/gangumales
Forest elephants: http://tinyurl.com/lukuruelephants

Based on the results of our surveys, which they helped to finance, African Wildlife Foundation decided in 2015 to fund a team of wildlife guards to patrol the Bili-Gangu Forests. You can read more about their project here: http://www.awf.org/projects/bili-uele-chimp-survey

The Discovery of the First Conodont Animals

Euan N K Clarkson

Euan graduated with an MA in Geology from Cambridge University, then as a PhD student, working on vision and other aspects of trilobite biology. He obtained a lectureship at Edinburgh University, later rising to Professor of Palaeontology and retiring in 2002. His textbook Invertebrate Palaeontology and Evolution *(Blackwell science, 4th edition 1998) is used globally, and he has written other books and over 150 other research publications on trilobites and other fossil arthropods, regional geology and palaeoenvironments of south and central Scotland.*

Introduction - what are conodonts?

It is not uncommon for new, and important scientific discoveries to be made by accident, during an experimental or observational research programme which was set up for quite different purposes. This was very much the case with the discovery, in 1982, of the first known conodont animal. The history of the 150-year search for, and the first discovery of the this soft-bodied fossil has already been superbly documented by Knell (2013) and this present paper is not aimed at repeating this excellent work. It is simply a personal testament of what happened during the stirring weeks of 1982-3, and what came before and after. But before we record the history of this event, we should discuss what conodonts actually are.

Our knowledge of conodonts began with the Baltic German embryologist and palaeontologist Christian Heinrich Pander (1794-1865), a truly great scientist, particularly well-known as an embryologist. During the 1830s and 40s he was investigating the faunal contents of relatively soft fine siltstones from the marine Ordovician and Silurian of Estonia, and he found, in the washed residues, great numbers of tiny, lustrous, tooth-like fossils.

He called these Conodonten, and in 1856 he published a monograph of these, describing 56 species and 14 genera, all with different morphology. His observations were excellent, and his belief that the conodont elements, as they are now normally termed, were the teeth of an extinct group of fishes, proved much later to be not far short of the mark. He discussed, but was unable to conclude, whether the individual conodont animals had only one kind of conodont element, or whether there were several kinds within the mouth of the animal, in other words differentiated teeth, as in mammals. Since the original body must have been soft and normally incapable of preservation, as Pander clearly recognised, he was not prepared to speculate upon its nature. And just what the conodont animal actually was remained one of the great mysteries in all of palaeontology until 150 years later.

Following Pander, several palaeontologists embarked upon conodont research, and it became clear that conodonts could be found in fine siliciclastic and carbonate sediments ranging from the upper Cambrian to the late Triassic, from which they could be dissolved out with acid, in specialised laboratories. But it was not until the 1920s that the first glimmerings of their stratigraphic usefulness began to be evident, and this accelerated the pace of research exceedingly. And then independently in America (Scott, 1934) and Germany (Schmidt, 1934), bilaterally symmetrical clusters containing several kinds of conodont elements, with right- and left-handed forms, were discovered on the surfaces of Carboniferous black shale slabs. Presumably, when the conodont animals had died, the sea floor was stagnant or undisturbed by currents, the soft body rotted away, and the natural assemblages were preserved, more or less in place, and with the conodont elements in their original relationship.

Not everyone believed it at first, some reputable workers were scathing. There seem to have been several reasons for this. One was that Scott's assemblages contained different kinds of elements, but in all cases there had been some minor current-sorting so that none were in their original relationship. Might the association of different types therefore have been fortuitous? A reasonable question, and only resolved when many bilateral symmetrical assemblages were later discovered, and the evidence became unassailable. A further point made by Knell was that in the politically turbulent year of 1934, there was much anti-German sentiment in the USA, and this affected the reception of German science in that country. And there was also, as I believe, the spectre of scientific jealously, which regrettably is often present when important discoveries are made, despite the apparent rationality of scientists.

But with the discovery of more natural assemblages, the original conodont animal became recognised (e. g. Aldridge et al 1987, Knell, 2013) as evidently bilaterally symmetrical, with the natural assemblages containing different kinds of elements. It seemed likely that they had functioned as teeth, with the different elements specialised for separate functions. Yet here again, this was disputed by some, in the absence of further evidence, which could only come from the discovery of the soft body.

There were great debates also on the taxonomic questions raised by the recognition of different elements having come from the same original animal, but that is beyond the scope of the present article. Further work involved, for industry as well as academia, much refinement of conodont stratigraphy. Moreover, it was recognised that conodont elements changed colour when heated, from pale yellow to brown, then black, and finally transparently clear. There soon came a very practical application for this phenomenon in the oil industry. Shales and other source rocks require a certain amount of natural heat to mobilise the oil and to ensure its migration to an oil reservoir rock, such as a sandstone with plenty of pore spaces. There is thus an 'oil window', representing the right conditions for oil accumulation in a reservoir from which it can be extracted. In other words not too little heating and not too much. The use of this Colour Alteration Index can show which rocks have been heated too intensely to retain oil, and it has proved to be a godsend because of its simplicity and ease of use.

But what of the original conodont animal? Speculation was rampant in the 40s to the early 80s. Conodont elements were referred by various investigators as belonging to almost every possible phylum. For some they were teeth; they not uncommonly showed signs of wear. For others they could not possibly have been teeth. Various affinities were postulated for them, plants, annelids, nematodes, molluscs, lophophorates, chaetognaths, agnathans and jawed fish. But by the later 20th century it became increasingly clear that if the conodont animal was to be found at all, would it not be best to look amongst the increasing number of described Fossil-Lagerstätten? This term refers to fossil-bearing localities where much more palaeontological information than usual is preserved. It originated from the late Dolf Seilacher of Tuebingen, Germany, one of the world's most far-sighted and influential palaeontologists of any time. He distinguished Konservat-Lagerstaetten, in which much more information than usual is preserved in the fossils, as a result of unusual environmental chemistry, from Konzentrat-Lagerstaetten where the usual kinds of shells and bones are preserved, but much more abundantly. The former are clearly of greatest interest here.

The search for the original conodont animals, especially in the context of Konservat-Lagerstaetten, has been so eloquently described by Simon Knell (2013) that we do not need to go into it in detail. Suffice it to say that any new assignation was treated with great interest, if not necessarily with respect. A primary contender, at least for some years was Typhloesus, an enigmatic soft-bodied metazoan from the Bear Gulch Lagerstatte in Montana USA. First described by Melton and Scott (1969) and later colloquially referred to as the 'Beast of Bear Gulch (Knell, 2013) this curious, soft-bodied, headless, bag-shaped animal, possessed a substantial gut that was often replete with conodont elements. Whereas this seemed at first to be the real conodont animal, it was subsequently recognised (Conway Morris 1990) that some specimens contained a mixture of many kinds of conodont elements, others just a few or none at all. Moreover, natural assemblages were absent. Was not *Typhloesus*, therefore, a predator, whose diet included conodont animals? This became ever clearer, and eventually, though not without regret on behalf of many people, the concept of *Typhloesus* as a conodont animal was abandoned.

Another possibility was, a flattened, segmented slipper-shaped animal, Odontogriphus (http://tinyurl.com/odontogriphus), in the Middle Cambrian Burgess Shale Fossil-Lagerstatte of British Columbia. This, described by Simon Conway Morris (1976) had a double ring-shaped organ (lophophore) on its head, with little spikes arranged along its length. Each spike may have had a tentacle covering it. But were the small spikes conodonts? After an initial burst of enthusiasm for this new creature, it became increasingly clear that this was not, after all, the true conodont animal, and so the search went on.

Scottish Carboniferous Shrimp Beds

Scotland today has three main geological and topographical components, the rugged, and mainly metamorphosed Highland in the north, the highly populated Midland Valley, which consists of Palaeozoic sediments and igneous rocks, and the wild hill country of the Southern Uplands, made up mainly of highly folded and faulted Lower Palaeozoic sediments of deep marine origin. The Midland Valley is bounded to the north-west by the Highland Boundary Fault, and to the south-east by the Southern Upland Fault. During the Lower Carboniferous the basic structure was the same, but a great, fresh-or brackish lake extended over the eastern part of the Midland Valley, hemmed in by the boundary faults and by a marine delta in the east and a volcanic plateau in the west. The lake was ringed by forest, and abounded in fish and crustaceans; it was subject to occasional marine incursions from the east. The coastal delta-plain settings around the lake, often with algal mats, provided ideal conditions for preservation of soft-bodied animals, including the crustaceans, as did at least one thermally stratified lake, possibly isolated from the main water body. None of these 'shrimp-beds' was deposited in a fully marine environment, though most have some marine influence.

My own involvement with the shrimp-beds began in the early 1970s, some years after I was appointed to a lectureship at Edinburgh University in 1963. In our departmental collections, and in the National Museum of Scotland there were many specimens from these Lower Carboniferous shrimp-beds. There were two of these horizons locally, one with a shrimp known as *Tealliocaris* in East Lothian, to the east of Edinburgh, the other at Granton, within the city limits, on the north shore. These shrimps had been ably described by Ben Peach (1908), in a fine monograph of Carboniferous crustaceans, but it seemed to me that there was much more to be done, using techniques unavailable to Peach.

Although slabs of the Granton material had been illustrated (Tait 1917) and the common form *Waterstonella* had been described and named by Fred Schram (1979) then based in Chicago, there was again scope for further researches, especially since the Granton fauna, though dominated by the small *Waterstonella* was to prove much more diverse.

The Gullane locality was not very easy to locate. My old Professor, the late Sir Frederick Stewart, had visited it previously and told me that it lay at the western end of Cheese Bay. Apparently, at some time in the 17th or 18th century a French ship, laden with cheese, intended to be sold to the gourmets of Edinburgh, had come to grief on the shore during a storm. The local population dined well on cheese for several weeks as a result. The 10 cm bed containing the fossils was normally buried by sand, though very many blocks of crustacean-bearing material could be found amongst the sea-weathered boulders along the shore. Sir Frederick, however, had told me that the shrimp-bed itself was sometimes exposed in February and March after the winter storms had removed the sand. This indeed proved to be the case.

My colleagues from Aberdeen University, had become very interested in the laminated sediments yielding the shrimp, and it was possible to measure the section (Hesselbo & Trewin 1984) and interpret it as recording the deposits of a thermally stratified, probably freshwater lake. The Granton section was less of a problem. On Saturday and Sunday mornings, when my then adolescent elder sons played rugby football at a club quite close to the shore, I left them there and went exploring along the shore, trying to find Tait's original locality. Again, there were quite a number of shrimp-yielding boulders littered along the shore, which yielded good specimens. There was also a series of calcareous and shaly horizons, slightly folded and well-exposed. Searching through these exposed sediments, I found three separate shrimp-bearing horizons, each probably representing a mass-mortality event. This was, finally the 'mother-lode"!

By the mid 70s, I had begun to make preliminary sketches, both of the Gullane and Granton faunas. At the December meeting of 1976, in Reading, of the Palaeontological Association, I had discussed these faunas with Derek Briggs, then of Goldsmith's College, in London, and asked if he would care to join me in their description. He was happy to do so, but because of other commitments we could not really commence until 1980. And so the work began, starting with *Tealliocaris.*

Firstly, I obtained a chemical analysis of the white mineral which so delicately had picked out all the details of structure, and it proved to be fluorapatite, $Ca_5(PO_4)_3F$ of which more anon. Secondly I went, for a week, to the Dunstaffnage marine laboratory, near Oban, western Scotland, in order to become thoroughly familiar with the anatomy of living shrimps and other eumalacostracan crustaceans. Returning to Edinburgh, I began to explore the anatomy of *Tealliocaris*, and the Granton faunas, making 80 and more drawings using a camera-lucida microscope, and taking photographs with the light and scanning electron microscopes.

The Carboniferous crustaceans were remarkably similar in anatomy to the shrimps of today, apart from the fact that none had claws or pincers. Evidently shrimps are a fine example of 'good design', which has persisted for at least 350 million years. I had quite a lot of material which I had collected from both localities and I had studied all the type and other specimens of *Tealliocaris* from the Geological Survey's collections, and those of the National Museum of Scotland. Work progressed well, but we had to abandon it for the time being so that we could produce a paper on the Granton fauna for the 'Trilobites and other early arthropods volume' Special Paper 30 (Palaeontological Association,Briggs & Clarkson 1983).

In the course of this work, in February 1982, I examined the collections from Granton in the Geological Survey, at Murchison House in Edinburgh. The then curator, Peter Brand, was most helpful. He had already laid out all the specimens on a large table and as we were going through them he said,"Here's a strange creature, a fish of some kind. Both part and counterpart." It looked interesting. Clearly it was something other than a shrimp. So, I borrowed the specimens from the Survey on a Friday afternoon in February, and after the registration process was completed, collected them the following Monday morning, and studied the fish-like creature with a binocular microscope. My first impression was that it looked like a tiny lamprey, especially the thin ray-supported fins which defined the tail end.

It was preserved, like the shrimps, in fluorapatite, but also, in some parts, in a blue mineral known as vivianite. But what was this? Within the head end there were small, spiny, black objects. They showed more clearly when water was applied to the surface. They did not seem very similar to the horny, peg-like teeth of living lampreys, and it crossed my mind that they might be conodonts, but at first I could not believe that this is what the black objects might be. So that evening I phoned Derek Briggs, and told him that I thought we had a strange fossil. He was bathing a squalling baby at the time, but I managed to get the message across. A few days later I went down to London to see him, bringing the specimen with me. By this stage I had re-examined the animal several times, and had once more begun to wonder if the spiny black objects could be conodonts. But I kept quiet about this; I wanted to hear what Derek had to say about them. We poured water on the head to bring out the details.

"Surely those are conodonts!" said Derek. "You don't suppose we've found the conodont animal?"

I tried hard to be skeptical, but this was proving all too exciting. Derek had to go into the city that afternoon, while I remained behind, making camera-lucida drawings of the creature. Later that afternoon he phoned,"Dick Aldridge will be along as soon as he can make it."

Dick was a Reader, later Professor at the University of Leicester, well known for his researches on British Silurian conodonts as well as those from other countries. He duly arrived, and immediately confirmed, on the basis of his considerable expertise, not only that the objects in the head were conodonts, but that they appeared to form part of a natural assemblage.

"This is the most interesting occurrence of conodonts, in the most interesting situation, that I have ever seen."

He smiled seraphically and went on his way. By the time he was home, as he told me later, he was convinced that it was actually the fabled conodont animal. We three prepared our paper for 'Lethaia', (Briggs et al., 1983), and presented our results at the annual meeting of the Palaeontological Association. My good friend Keith Ingham, at Glasgow University had a superior photographic apparatus, and he took fine pictures of the conodont elements in their preserved relationship. We illustrated also a fused cluster of conodont elements, isolated by acid, and it may be that we had unwittingly dissolved another one of the conodont animals.

There were remarkably few skeptics, and the conodont specialists particularly were very kind and encouraging. We were very happy about the reception of our paper. Somebody had to find conodont animals eventually, and quite by chance it happened to be our small team. But whoever had made the discovery, and thereby had solved one problem, had opened new perspectives on conodont animal biology and relationships, which was ably undertaken later by the Leicester University group, headed by Dick Aldridge.

Derek Briggs and I obtained funding for our researches on the Carboniferous shrimp beds of Scotland, including several more in the eastern Scottish Borders, and in 1985 we were able to take on Neil Clark as our main field man, who was to locate further shrimp beds. He continued to make very fine studies on crustaceans (e. g. Clark 2013) after Briggs and I had concluded our researches. Moreover we took on John Cater as our sedimentologist, and he was able to make great progress in interpreting the environments (Cater 1987, Cater *et al.* , 1989, Briggs & Clarkson 1989, 1991).

The most likely scenario for the environment (though there are alternatives) is that the Granton shrimps, and various kinds of fish formed an indigenous community within the lake, which may have been fresh or brackish water at the time. The eastern end of the lake was close to the sea, and subject to periodic invasion by the sea during storm surges. These brought in marine organisms such as straight-shelled nautiloids, and also the soft-bodied conodont animals and tomopterid worms, all of which appear slightly rotted, as if they had travelled some distance. The drastic change in salinity killed the shrimps, hence the mass mortality horizons. Whereas the phosphate was derived from marine phytoplankton blooms, it was the rapid spread of coccoid phosphate bacteria over the surfaces of the dead invertebrates that preserved them. What we see in the fossils, in which every detail is picked out, is the remains of these bacteria. It is the same with *Tealliocaris*at Cheese Bay (Briggs & Clarkson, 1985) though there are no distinct mass-mortality levels in the sequence, and these beds were probably deposited in an isolated, thermally stratified fresh-water lake on the outside the main water body.

It was Neil Clark who found the second conodont animal in the Granton Shrimp Bed, a couple of days after he had finished his final year undergraduate examinations at Edinburgh University, in June 1983. It was to prove to belong to a different species. This, and another two specimens which he discovered soon afterwards, were described in a second Lethaia paper (Aldridge *et al.* 1986). I went on holiday in June and almost immediately after I arrived home there was a phone call from Neil. Some private collector or dealer had ripped out several square metres of the Shrimp Bed. To prevent further damage I called the then Nature Conservancy Council, after discussing with the Survey and Museum authorities, and Edinburgh City Council. They sent a heavy lorry and a gang of strong men. The Shrimp Bed was removed, and taken to the National Museum's storage unit. In one sense, despite the destruction, this was a blessing. For in the course of studying this material several new crustacean species were found. But more importantly, there were no less than six new specimens of conodont animals, which makes ten from Granton altogether. We discovered the first known fossil tomopterid worm (Briggs & Clarkson 1987b), of which further specimens were found by Neil Clark, and a previously unknown chordate, *Conopiscius* (Briggs & Clarkson, 1987a). Though the bulk of our work is concluded, other creatures remain to be described. A possible sipunculid has recently been described (Botting & Muir, 2007) and ubiquitous branching organisms, possibly hydroids (or are they sponges or algae) are under description. Might other conodont animals from here turn up? We live in hope.

My own contribution to the discovery of conodont animals, such as it was, had now finished. Derek Briggs at Bristol, and his colleagues continued with studies of preservation. Further interpretation of conodont animals was undertaken by the vigorous and active research group at Leicester University, headed by my late friend, Dick Aldridge, Paul Smith, Mark Purnell, Phil Donoghue, Ivan Sansom, Sarah Gabbott and others, and their colleagues elsewhere were highly experienced. They were the obvious scientists to take the research work forward. I returned to my studies of trilobites.

Meanwhile, now that it was known what the conodont animal looked like it was possible to tie up many loose ends. Firstly, the Granton conodonts, eight of which are of the genus *Clydagnathus* are elongated and up to 55mm in length.

They have a short head, a trunk with v-shaped myomeres (muscle bands) and a ray-supported caudal fin (remarkably like that of a lamprey). The position of a notochord is represented by a pair of parallel lines extending along the dorsal part of the body. A pair of large capsules at the anterior end represent the eyes (Aldridge *et al.,* 1993, Purnell, 1995). There is little doubt that the basic morphology of the conodont animal is that of a chordate, most likely a vertebrate (and obviously a predator); but the debate on the precise affinities thereof rumbles on. It has also been possible to re-study the natural assemblages of conodont elements, and how they worked. The Granton animals, mostly with an 'ozarkodinid' type assemblage at least showed which end was which! The anterior part consisted of a paired series of 'hindeodellids' , or S elements, each an elongated comb of sharp needle-like teeth and a more curving but otherwise fairly similar M element at each side. This part of the assemblage was highly inclined to the body in life, the paired elements facing inwards as a formidable predatorial apparatus. Behind these were the Pb elements, stout, curved and with denticles and a central cusp, and posterior to these again, a pair of strong massive Pa elements. In a sense their function can be distantly compared with that of our own dentition of incisors, canines and molars. A fine example of convergent evolution. Meanwhile the debate on conodont affinities continues (Aldridge & Purnell 1996, Smith,. 1990).

The discovery of a poorly preserved Silurian panderodontid conodont from Wisconsin was a further step forward, as was the finding of exceptionally large *Promissum* in the Ordovician Soom shale of South Africa, a giant conodont animal, originally described as a plant (Gabbott *et al.* 1995). So, there the matter rests. We really need to find more conodont animals belonging to other groups. And I am certain, now that we know what to look for, that we shall surely do.

Acknowledgements

I am deeply grateful to Professor Brigitte Schoenemann (Cologne) and Dr Tom Challands (Edinburgh) for reading this paper and for helpful discussion.

References

This is by no means a comprehensive list of literature pertaining to conodonts; it concerns only those papers relevant to the discovery of conodont animals.

Aldridge, R. J. (ed.) 1987. *Palaeobiology of conodonts*. British Micropalaeontological Society Series. Ellis Horwood, Chichester. 180pp.

Aldridge, R. J., Briggs, D E. G, Clarkson, E. N. K. & Smith, M. P. 1986. The affinities of conodonts - new evidence from the Carboniferous of Edinburgh, Scotland, *Lethaia*, 19, 279-291

Aldridge, R. J., Briggs, D. E. G., Smith, M. P., Clarkson, E. N. K., and Clark, N. D. L. . 1993. The anatomy of conodonts. *Philosophical Transactions of the Royal Society of London* B 340, 405-421

Aldridge, R. J. and Purnell, M. A. 1996. The conodont controversies. *TREE* 11, 463-468

Briggs, D. E. G. & Clarkson, E. N. K. 1983. The Lower Carboniferous Granton 'shrimp-bed', Edinburgh. *Special papers in* Palaeontology 30, 20-22

Briggs, D. E. G, Clarkson, E. N. K. & Aldridge, R. J. 1983. The conodont animal. *Lethaia*. 16, 1-14

Briggs, D. E. G. & Clarkson, E. N. K. 1985. The Lower Carboniferous shrimp *Tealliocaris* from Gullane, East Lothian. *Transactions of the Royal Society of Edinburgh: Earth Sciences* 76, 35-40

Briggs, D. E. G. & Clarkson, E. N. K. 1987b. The first tomopterid, a polychaete from the Carboniferous of Scotland. *Lethaia*. 20, 257-262

Briggs, D. E. G, and Clarkson, E. N. K. 1987a. An enigmatic chordate from the Lower Carboniferous Granton 'shrimp-bed' of the Edinburgh district, Scotland. *Lethaia* 20, 107-115

Briggs, D. E. G. and Clarkson, E. N. K. 1989. Environmental controls on the taphonomy and distribution of Carboniferous malacostracan crustaceans. *Transactions of the Royal Society of Edinburgh; Earth Sciences* 80, 293-301

Briggs, D. E. G., Clark, N. D. L., and Clarkson, E. N. K. 1991. The Granton 'shrimp-bed', Edinburgh - A Lower Carboniferous Konzervat-Lagerstätte. *Transactions of the Royal Society of Ednburgh; Earth Sciences.* 82, 65-85

Cater, J. M. L. 1987. Sedimentology of part of the Lower Oil-Shale Group (Dinantian) sequence at Granton, Edinburgh, including the Granton “shrimp-bed". *Transactions of the Royal Society of Edinburgh* 78, 29-40

Cater, J. M. L., Briggs, D. E. G. and Clarkson, E. N. K. 1989. Shrimp-bearing sedimentary successions in the Lower Carboniferous (Dinantian) Cementstone anjd Oil-Shale Groups of Northern Britain. *Transactions of the Royal Society of Edinburgh; Earth Sciences*. 80, 5-15

Clark. N. D. L. 2013. *Tealliocaris*, a decapod crustacean from the Carboniferous of Scotland. *Palaeodiversity* 6, 107-133

Conway Morris. S. 1976. A new Cambrian lophophorate from the Burgess Shale of British Columbia. *Palaeontology* 19, 199-222

Conway Morris, S. 1990. *Typhloesus wellsi* (Melton & Scott, 1973), a bizarre metazoan from the Carboniferous of Montana, U. S. A. *Philosophical Transactions of the Royal Society of London* B, 595-624

Gabbott, S. E, Aldridge, R. J. and Theron, 1995. A giant conodont with preserved muscle tissue from the Ordovician of South Africa. *Nature*, 394, 800-803

Hesselbo , S. P. and Trewin, N. H. 1984. Deposition, diagenesis, and structures of the Cheese Bay shrimp bed, Lower Carboniferous, East Lothian. *Scottish Journal of Geology* 20, 281-296

Knell, S. J. 2013. *The Great Fossil Enigma; The Search for the Conodont Animal.* Indiana University Press, Bloomington & Indianapolis. 413pp

Lindström, M. 1964. *Conodonts.* Elsevier, Amsterdam. 196pp

Lindström M. 1974. The conodont apparatus as a food-gathering mechanism. *Palaeontology* 17, 729-744

Melton, W. G. and Scott, H. W. 1973. Conodont-bearing animals from the Bear Gulch Limestone, Montana. *Geological Society of America Special Paper* 141, 31-65.

Muir, L. A. & Botting, J. P. 2007. A lower Carboniferous sipunculan from the Granton Shrimp-bed, Edinburgh. *Scottish Journal of Geology* 43, 51-56

Pander, C. H. 1856. Monographie der fossilen Fische des silurischen Systems der russich-baltischen Gouvernements. *Koniglisches Akademie wissenschaften. St Petersburg.* 91pp

Scott, H. W. 1934. The zoological relationships of the conodonts. *Journal of Paleontology* 8, 448-455

Schmidt, H. 1934. Conodonten-Funde in ursprünglichen Zusammenhang. *Palaontologisches Zeitschrift*. 16, 76-85

Schram, F. 1979. British Carboniferous Malacostraca. *Fieldiana: Geology* 40, 1-129

Smith, M. P. 1990. The Conodonta - palaeobiology and evolutionary history of a major Palaeozoic chordate group. *Geological Magazine*. 127, 365-369.

Purnell, M. A. 1995. Large eyes and vision in conodonts. *Lethaia* 28, 187-188.

Purnell, M. A. 1995. Microwear on conodont elements and macrophagy in the first vertebrates. *Nature* 374, 798-800

Smith, M. P., Briggs, D. E. G. & Aldridge, R. j. 1987. A conodont animal from the Lower Silurian of Wisconsin, U. S. A., and the apparatus architecture of panderodontid conodonts. In *Palaeobiology of conodonts* (ed. R. J. Aldridge), 91-104. Ellis Horwood, Chichester.

Sweet, W. C. 1988. *The Conodonta ; morphology, taxonomy, palaeoecology and evolutionary history of a long-extinct animal phylum.* Oxford Monographs of Geology and Geophysics 10, 212pp

Tait, D. 1925. Notice of a shrimp-hearing limestone in the Calciferous Sandstone Series, at Granton, near Edinburgh. *Transactions of the Edinburgh Geological Society* 11, 131-135

Unraveling Collagen Biology Enhances Human Life, Health, and Beauty

James D. San Antonio and Olena Jacenko

James and Olena obtained Ph.D.s in Cell and Molecular Biology at the University of Pennsylvania and both did postdoctoral research at Harvard Medical School. James was an Associate Professor at Jefferson Medical School, a graduate course lecturer, and is now a scientist in Biotechnology. He has more than twenty years of experience in collagen and vascular biology and is an inventor of pharmaceuticals and medical devices. Olena has been a faculty at the University of Pennsylvania since 1994 and is now Professor of Biochemistry in the Department of Animal Biology, and Associate Dean of Faculty Affairs and Diversity at the School of Veterinary Medicine. She has more than twenty years of experience in collagen and skeletal biology, runs an extramurally-funded research laboratory, and teaches basic science to veterinary students. Olena has obtained numerous awards for excellence in research and teaching.

The bodies of humans and other vertebrates are made mainly of two things- cells that carry our genetic information and perform the many functions of life, and the "stuffing" between cells, called the extracellular matrix, or the matrix. In fact, the vertebrate body can be viewed as a scaffold made up of an intricate network of matrix molecules, whose presence dictates the mechanical structure and function of a particular tissue or organ. Cells live within this matrix scaffold, and obtain from it both structural support and instructional cues as to how they should behave. The protein collagen is by far the most abundant of all the matrix molecules in vertebrates. Without this internal architecture composed of the collagen-rich matrix, you- our dear reader- would be a shapeless pudding of cells on the ground, haplessly crawling about like an amoeba!

Collagen makes up much of bones, cartilage, tendons, ligaments, skin, teeth and other connective tissues. Collagens are also present in lower organisms including worms, jellyfish, flies and many others– where they also play crucial supportive roles. In the human body, there are at least thirty types of collagen. By definition, all collagens must meet three requirements. First, they have to be mostly composed of triple helical, three stranded, protein chains that fold into tight rope–like molecules.

Second, these trimeric molecules must assemble into larger structures called aggregates, either with themselves or with other matrix molecules. Third, they have to be secreted outside the cells where they form the intricate networks that support the structure and functions of the cells and the whole organism.

The most abundant of all matrix molecules in humans and other vertebrates is type I collagen, which assembles and twists together into molecules that look like massive cables. These collagen cables are further glued together, via chemical crosslinks or "molecular welds," giving them incredible strength. Thus, gram for gram, collagen is stronger than steel! Interestingly, some collagen fibrils closely resemble concrete reinforcement bars– or rebar– which are steel rods recognizable to most people who have glanced upon a building construction site.

Rebar is embedded within concrete to increase its tensile strength and its surface is irregular to lessen the chance of slippage between it and the surrounding concrete when exposed to external forces. So too, the surface of collagen fibrils is irregular, likely increasing its hold on and thus support of the surrounding tissues of the human body.

There is approximately 1×10^{21} type I collagen molecules, or about 7 kilograms of the protein in the human adult. If these type I collagen "ropes" from one adult were unraveled and the individual collagen molecules laid end to end, they would span farther than the distance from the earth to the sun! The abundance of collagen is noteworthy also as a key factor in human life, nutrition, and medicine.

The word collagen is derived from the Greek "Kolla" (glue) and the French "gene/gen" (producer of). Throughout history the collagen from animal skins and bones has been used for the manufacture of glue, and makes up the bulk of leather used for tools, clothing, and coverings for household or automobile seats. Collagen–rich tissues such as cow or sheep intestines or reconstituted collagen tubes are used to manufacture strings for musical instruments or sports racquets, and sausage casings. Collagen from skin and bones is boiled and hydrolyzed– or broken apart– to produce gelatin, an important component of many foods, e.g, jello, aspic, and confections, and is added to foods and pharmaceuticals as a thickener. It is also used to fabricate drug capsules or as a settling agent in the brewing of beer.

The past century has brought huge leaps in science in general as well as in the collagen field. Development of new techniques allowed scientists to progress immensely from a vague understanding of collagen in the early 1900's as a ubiquitous, poorly defined structural material in the human body. As early as the 1950's and 60's, the newly invented electron microscope helped visualize collagen at high resolution as highly organized, rope– or cable–like structures, and prompted chemists to dissect its amino acid composition and sequence. At that time, some of science's biggest names including Linus Pauling, James Watson, and Francis Crick raced to elucidate collagen's complex three dimensional shape while also deciphering that of DNA. Ultimately, however, it was G.N. Ramachandran who solved the basic structure of collagen. Ramachandran and his co-workers used X-ray diffraction to examine samples of collagen from kangaroo tail tendon-where the molecules are relatively well aligned with each other. In this approach the electrons of the atoms making up collagen diffract or move the X-ray beam giving clues as to what type of atom they are, which other atoms they are associated with, as well as, their precise positions in the amino acids that make up the collagen chains. Ramachandran used this information to model, and thus discover how the collagen chains are arranged as a three-stranded bundle, or triple helix, unlike the double helix making up DNA or the alpha helix commonly found in many other proteins.

Thereafter scientists discovered that several other structural molecules routinely associate with collagen to build tissues and made great progress delineating how cells secrete and assemble collagen. A crucial discovery on how collagen is broken down for removal by the body was made by Jerome Gross and Charles Lapiere. These scientists noticed that collagen–rich tails of tadpoles disappear as they develop into frogs and asked the question– is there a factor in the tail that degrades collagen?

They removed tadpole tails and lay them onto beds of collagen in petri dishes– and sure enough, after some time had passed, they noticed a clear halo forming in the collagen around the tails, indicating that collagen was being degraded. The active factor– an enzyme or "molecular scissors"– called collagenase– was later isolated. Since then, other collagenases have been discovered, and this class of enzymes has been shown important in human diseases like arthritis.

In the last few decades the proof that mutations in collagen may cause serious human diseases like osteogenesis imperfecta (brittle bone disease) was determined; collagens were cloned and expressed in cell cultures and plants, and genetically modified transgenic mice (experimentally manipulated mice that carry foreign genetic information) as models of human disease were created, leading the way to the identification of a huge family of collagens and a detailed understanding of the genetic basis of many human diseases.

There are many other modern applications of collagen technology– some are currently in development and others are now in use. They include for example, using enzymes (including collagenases) that break down collagens to selectively eliminate the collagen component of scars in skin, or within fibrotic lesions in livers after infection or alcoholism, or life–threatening adhesions formed between organs or tissues after some surgeries. Other approaches aim to block collagen assembly or scar formation in such diseases using small drugs designed to interfere with the aggregation and growth of the collagen molecules. Various collagen formulations are used to manufacture biocompatible and biodegradable medical devices to repair or replace diseased or injured tissues, including bone, tendons, and ligaments. Collagen is often a key component of hemostats, which are devices designed to stop bleeding, for example, during surgery. Collagen binds and activates platelets which are crucial clotting factors in blood, and strengthens blood clots. Collagen also strongly promotes the growth and regeneration of small blood vessels. Healthy collagen is important for human beauty and a youthful appearance, as the aging of collagen can lead to skin sagging and wrinkling. Therefore, collagen is big business in the beauty arena! For cosmetic applications, collagen fillers are injected under the skin to smooth wrinkles or fill voids caused by injury, disease, or aging, and are used for breast and other tissue implants. Collagen is a common ingredient in skin creams and shampoos, where it imparts hydrating or water binding properties, and can help diminish the appearance of wrinkles.

For the above applications collagen is often isolated from animal skins or bones, but alternative approaches in development aim to use recombinant DNA technology to express collagens in tobacco, corn, or other organisms where they normally never exist. Benefits of the latter include high volume, low cost manufacturing, eliminating the risk of pathogens that may co−purify with collagen when it is isolated from mammalian sources like cattle skin (such as the causative agent of mad cow disease), alleviating the need to use animals in the production process, and the ability to manufacture "designer" collagens with desirable attributes. In general, such collagens would be engineered to be non-immunogenic, or non−reactive with the human body. Collagens would be further modified for specific applications. To help repair heart tissues damaged by a heart attack, one might want to deliver a collagen endowed with a superior ability to help blood vessels regenerate. On the other hand, a collagen destined for injection to plump lips or smooth skin wrinkles should be engineered to bind maximal amounts of water and to last longer in the body, requiring less frequent injections.

Collagen Research and Applications

There are many great questions or scientific holy grails remaining to be answered in the collagen field, some of these include: how is the collagen molecule assembled, and which parts are on the outside and which are on the inside? Which parts of collagen help mineral to grow and form bone? How do mutations in collagen lead to human disease, and can such diseases be effectively treated or even cured? How can the structure and function of the many complex types of collagen be solved and understood?

Type I collagen road map

As a graduate student one of us (James) attended a research seminar given by Merton Bernfield, an eminent Professor at Harvard Medical School. He presented discoveries on the importance of epithelial (skin) cell interactions with collagen, which is relevant to diseases like cancer. After the lecture I asked Merton "to which part of the collagen molecule do skin cells bind?" He responded that figuring out where cells and various molecules interact with collagen is a significant challenge and that no one had made much progress in that regard.

That encounter inspired my co-workers and me to devote more than a decade to performing experiments that discovered where skin and blood vessel cells bind to collagen, as well as how certain other collagen-binding molecules interact. Luckily, other researchers shared our passion so by the end of the twentieth century many dozens of binding sites on collagen had been determined, a few hundred collagen mutations associated with human diseases were mapped, and additional knowledge about collagen structure and assembly had also been elucidated by structural biologists.

In spite of this wealth of information, no database or "road map" of the protein existed showing its structure and repertoire of functions and mutations on one page. Thus, consider a typical map of a small city, where restaurants, businesses, city hall, and schools might concentrate to the center of town and more utilitarian sites like water works, power plants, and the garbage dump locate to the outskirts. In this scenario, the distribution of sites in the city is non−random, and the map would show an outsider where the various aspects of life are accomplished. Similarly, the point of creating a collagen road map was to see if collagen is organized in any particular way, and to use this information as a guide to asking research questions about collagen biology.

So about fifteen years ago our lab began constructing a collagen map. We started by drawing a schematic of the basic collagen structure on large poster boards with sites of interest- which we found through our own research or which were discovered by other labs- marked in pencil and colored clear cellophane overlays tacked onto it to denote broad regions where larger molecules were known to bind. Of particular importance was our collaboration with a team of human mutation mappers led by Joan Marini, who provided us with the long list of human mutations they had painstakingly mapped to collagen. As the volume of data in the collagen literature grew, we worked with Drew Likens, a graphic artist, who digitized the data, and devised a color scheme to enhance its contrast, readability, and attractiveness. Since then the map has been revised dozens of times with new data, and now contains thousands of bits of information.

What does the map teach us? The simplest observations were that some molecular binding sites or structural features are near neighbors on collagen, and thus it logically follows that they may be functionally related. For example, collagen's cell binding sites were found to overlap with binding sites for various collagen-binding proteins, leading us to speculate that those collagen-binding proteins may influence cell interactions with collagen. Also, it was very clear that human mutations are distributed non-randomly on collagen. For example, there are some regions of collagen with no reported mutations, and other regions have clusters of either mild or severe mutations. We speculated the latter are signposts for functionally crucial parts of collagen. But does the map teach us any more significant lessons?

It took about five years of map building and studying the data before we discovered that collagen appeared to be organized in a simple way- it has two major parts, or domains. Thus, the dynamic aspects of collagen biology– where cells and most bioactive factors bind and where collagen can be broken apart by collagenase– are in one region, that was separate from another region where collagen fibrils are crosslinked or welded together, other molecules interact to form the matrix scaffold, and biologic mineral attaches in hard tissues like bone.

To gain support for our hypothesis required collaborating with many other scientists with different talents to help analyze and interpret our data. For example, we worked with Joseph Orgel, a structural biologist, who confirmed our hypothesis by using his 3D model of type I collagen derived from X-ray diffraction studies of collagen-rich rat tendon. We also collaborated with the statistician John McAuliffe, who helped us prove that our data analysis was rigorous enough to support our hypothesis. Finally, we discovered that other fibrillar collagens had a similar domain structure to that of type I collagen, which further bolstered our hypothesis.

Our discovery may have far-reaching implications. It should help geneticists predict which collagen mutations are the most harmful to human health, and possibly use this information to help lessen, or eliminate the effects of such mutations. Moreover, our discovery lays the groundwork for the rational design of new collagens with crucial applications in tissue engineering and human medicine, an approach discussed previously in this article. Thus, our domain model could teach the genetic engineer which parts of collagen may be modified to elicit new, exciting and useful properties, and which must remain untouched owing to their crucial biological functions.

Finally, it is worth mentioning that our collagen mapping studies met with substantial resistance from the scientific community. Many of our colleagues didn't feel our work to hold promise, granting agencies refused to fund the project, and we found it extremely difficult to publish our discoveries. However, at the same time, some of our colleagues were very supportive of the project and even used our map to help guide their research. Ultimately, the success of our collagen mapping project is reflected by the over four hundred citations our mapping papers have received in the scientific literature. Our struggles, which are probably not that unusual for scientists, reinforce the notion that adversity must be overcome to ensure progress and discovery.

The hunt for dinosaur collagen

Recently, we were delighted to find an unusual use for our collagen map. This story begins when the paleontologist Mary Schweitzer, while handling a well-preserved dinosaur bone noticed fibrous material poking out from some cracks in the fossil.

Mary knew collagen makes up most of the fibrous material in bones of living animals, so she dared ask– could the fibrous stuff be collagen that survived over 60 million years? Current dogma suggests that proteins like collagen shouldn't last very long after an organism dies– they should fall victim to decomposition by the action of bacteria and molds, and physical decay from forces such as freeze–thaw cycles and radiation damage. Much like if you left your home unoccupied for millions of years– would anything remain after bacteria, mold, termites, earthquakes, and floods had their way? But Mary and her colleagues looked for collagen in dinosaur bones anyway, using a relatively new technique called mass spectrometry that can identify little pieces of proteins. And amazingly– they found more than 10 small bits of the type I collagen molecule!

The fragments had sequences– or chemical fingerprints a lot like that of the collagen of living animals– implying the protein was relatively unchanged through the ages. This is as expected for such an important protein. However, most interestingly the dino collagen was more like that from modern birds, not mammals, supporting the dino–to–bird evolutionary argument. This led to a lot of colorful articles in the press, such as: "did T–Rex taste like chicken?" After reading Mary's papers we asked to work with her, using our collagen map to determine whether the pieces of dino collagen that she found locate to functionally interesting parts of the protein. Together we searched our human collagen map for sequences that looked highly similar or identical to those of the dinosaur collagen, and highlighted their positions on the human collagen map. We worked with a statistician- Shane Jensen- who is known in the popular sports literature for his statistical analysis of the performance of professional baseball players including Derek Jeter of the Yankees. For our work, Shane used a computer program to simulate how random distributions of collagen peptides would appear on our collagen map, and found that the patterns of peptides that survived in the dino bone could not have occurred by chance alone, suggesting that there must be a physical or biological reason those peptides persisted. We also found that some of the peptides located to the most crucial parts of collagen where cells and collagenase bind. Working with Joseph Orgel we then determined that on the living protein, the peptides would be localized to its inside, more protected zone, clueing us in as to why they may have survived.

Genetically engineered mice help decipher the role of a "stubby" collagen

Of the many collagens of vertebrates is type X collagen, considered a short–chain collagen. Although type X collagen appears stubby as compared with the long, slender type I collagen molecule, it also is barbell–shaped, having ball–like globular ends that associate end–to–end into a hexagonal–like network reminiscent of a woven basket or honeycomb. This protein forms such networks around cartilage cells – which, along with the other matrix molecules they secrete - make up the articulating joints between elements of the skeleton, and are also responsible for the growth of long bones. Two types of clues help us understand possible functions of type X collagen. First, humans who have mutations in type X collagen have a unique form of dwarfism called Metaphyseal Chondrodysplasia Type Schmid. Our own research probed the function of type X collagen in the mouse, by designing transgenic mice that have both reduced and abnormal assembly of the type X collagen scaffold. Interestingly, as in the humans, the mice are somewhat dwarfed, but surprisingly, also have problems with their bone marrow and the development of blood cells of the immune system. As a result, these mice cannot respond adequately when faced with an infection. We therefore propose that type X collagen plays crucial roles in skeletal growth and acts as a gatekeeper coordinating the flow of information and thus behaviors of the skeletal, blood, and immune cells that develop within the bone marrow.

Human Diseases Caused by Mutations in Collagen Genes

There are numerous significant human diseases caused by mutations in genes coding for collagens and other matrix molecules. For example, type I collagen mutations lead to brittle bone disease. The challenges of living with this disorder are poignantly described by Jane Hash, health professional and blogger.

Type II collagen defects are associated with many developmental disorders of cartilage (e.g. chondrodysplasias), such as osteoarthritis and dwarfism; these can range in severity from mild to debilitating to lethal. Mutations in type III collagen lead to Ehlers Danlos Syndrome, associated with highly stretchable, elastic skin and joints, and some vascular problems. Type IV collagen mutations often are associated with disorders of kidney function, while type VII collagen mutations cause blistering skin disease (Epidermolysis Bulosa). Individually, these diseases are considered moderately rare– affecting one in 10,000 individuals or less. There are many scientists and clinicians who specialize in understanding the molecular mechanisms causing these diseases, and work to find cures and improve the quality of life for those affected. The creation of transgenic mice and other animals having such diseases is emerging as a one of the most powerful tools towards understanding the mechanisms of these diseases. Moreover, understanding a disease mechanism, or the biochemical basis of one collagen disorder, can often be applicable to disorders involving other collagen types, and is an important step towards developing treatments for these diseases.

Collagen Research Culture

We are among the hundreds of collagen biologists and chemists around the world who study the structure and functions of collagens and their uses in industry and medicine. We publish our research in scientific journals and are members of societies such as the American Society for Matrix Biology, International Society of Matrix Biology, Osteogenesis Imperfecta Society, and Pan Pacific Matrix Biology Society. Although our family of scientists is large, many of us know each other personally or by reputation, and most can have their scientific lineage traced back directly or via one to three generations to a handful of pioneers in collagen biology. Thus on the family tree of collagen scientists, those comprising the trunk are some of the first individuals to study collagen structure and function in detail, whereas the fourth to fifth generation individuals are graduate students and trainees in the field. We– the authors of this article– are third generation, or small branches on the collagen biology tree. Our graduate school mentors and post-doctoral advisors, including Rocky Tuan, Henry Slayter, and Bjorn Olsen, did seminal work discovering or visualizing collagens for the first time, or themselves worked with individuals who did– including Jerome Gross, Cecil Hall, and Darwin Prockop. Our trainees are fourth or fifth generation twigs, and sadly, many are being blown off the tree by the stormy economy.

How healthy is the business of collagen science? Not very! A healthy research environment requires a lot of commitment and funding, and productive researchers require investment, cultivation, freedom to explore, and perhaps, most importantly, the means to maintain continuity in their research programs. Stagnating funding is leading to a generational gap– many of our trainees in recent years left academia, and, having emerged into a withering job market for research scientists, have reduced opportunities to innovate and explore in the collagen field. Most perniciously, this results in a lack of research continuity, with fewer scientists remaining to train new scientists and pass on the collagen culture. Without a robust research enterprise new scientists and teachers will have a greater challenge acquiring in–depth knowledge of their field of interest, and, as a result, the pace of fundamental discoveries in the collagen field has been diminishing, and will continue to do so. We hope the depressed economies in the United States, European Union, and elsewhere improve and the burgeoning economies in some other nations may take up the some of the slack to keep the business of collagen research and discovery chugging along.

In summary, collagens are the foundation of the human body and are needed for our everyday life, health, longevity, and beauty. Thus, continued research into the biology and practical applications of collagens will lead to improvements in the quality of human life for years to come.

The Life Work of a Sexual Health Educator and Researcher

Beverly Whipple

Dr. Beverly Whipple is a certified sexuality educator, sexuality counselor, and sex researcher. She is a Professor Emerita at Rutgers University in New Jersey and has co-authored seven books, one of which has been published in 22 languages, and authored or co-authored over 200 research articles and book chapters. She has received over 115 awards for her research contributions, including the Gold Medal from the World Association for Sexual Health and having a Scientific Research Center named after her in Puebla, Mexico "El Centro de Investigación en Sexología: Dra. Beverly Whipple." In addition, she was named one of the 50 most influential scientists in the world by the New Scientist for their 50th anniversary, one of only five women and the only person in the area of sexual health.

I was born in New Jersey in 1941 and have always lived in New Jersey. I received a BS degree with a major in Nursing from Wagner College in Staten Island, New York. My husband Jim and I were married in 1962 after I finished my undergraduate degree. We have two children, Allen, born in 1964 and Susan born in 1967, and 5 grandchildren. I worked as a nurse and a nursing instructor before I began my research concerning women's sexual health. As a nursing instructor I had to counsel students, and since I did not know anything about counseling, I studied for a Masters Degree in Counseling from Rutgers University, which I received in 1965. I took the courses for this degree before our children were born and when our son was an infant.

Sexual Health Educator

We have to go back to the early 1970's when I was teaching nursing in Camden, New Jersey. One of my students asked me, "What can a man do sexually after he has a heart attack?" I thought it was an excellent question, but I did not know the answer. I knew about climbing two flights of stairs without having shortness of breath, but that was all I knew. I spoke with the other nursing faculty and we decided to have a consultant from Marriage Council in Philadelphia help us to incorporate sexuality into our nursing curriculum. We worked hard to do this and then presented the new curriculum to the Board of Trustees of the nursing program for their approval. They said we could not implement the curriculum, because we would be talking about (listen to the word) "masturbUtion" and all those awful things.

So I quit my job, moved to another college's nursing program, and took three graduate courses in sexuality, one each summer, which were offered by the American Association of Sexuality Educators, Counselors and Therapists (AASECT). I started to attend professional sexuality organization meetings. At an EAST (Eastern Association of Sex Therapists) meeting in Philadelphia (EAST later became SSTAR [Society for Sex Therapy and Research]) I met Bob Franceour and many other well known sexual health professionals, including Alan Warbeck, Sandra Leiblum, Alex Comfort, and Wardell Pomeroy. Dr. Franceour invited me to be a facilitator at a SAR (Sexual Attitude Reassessment) he was giving at Farleigh Dickenson University in New Jersey and after this, I became more convinced of the need for sexuality education in the Nursing curriculum. Men and women who experience illness, medical treatments, or surgical interventions have questions about sexual health that were not being addressed or answered by the medical and nursing community. I started a sexuality course at my new college open to all students, not just nursing students. It was one of the most popular courses on campus. I also taught sexual health to nursing students, and was invited to speak to many medical students and physicians around the country. By this time, I was certified by AASECT as a sexuality educator and a sexuality counselor.

In 1980 I wrote an article about incorporating sexuality education for health professionals at the invitation of Dr. William Stayton, who taught one of the AASECT graduate courses that I took.[1]

I still give talks about sexual health to the general public and to professional groups world-wide. I have been blessed to have spoken in 94 countries to date.

Research concerning women's sexual health

I did not plan to become a researcher. At an AASECT meeting around 1977 I met Dr. John Perry, who had developed an electronic perineometer to measure pelvic muscle strength, which could also be used as a biofeedback device. I thought this device would be excellent to use to help teach Kegel exercises correctly to women who had stress urinary incontinence (SUI). This is when a woman loses urine when she coughs, jumps, or sneezes. So I began teaching Kegel exercises to treat stress urinary incontinence using the electronic perineometer to give the women feedback on how well they were doing the exercises. These exercises are named after Dr. Arnold Kegel, who found that the exercises helped prevent surgery in women with SUI.[2, 3]

Some of the women who came to me to learn the Kegel exercises for their stress incontinence, had very strong pelvic muscles, where with stress incontinence, the pelvic muscles are very weak. These women with the strong muscles stated that the expulsion of fluid from their urethra was triggered by stimulation of a sensitive area felt through the anterior or front vaginal wall. They also reported that this fluid from their urethra did not look or smell like urine, was about a teaspoon in volume, and tasted sweet.

A search of the literature lead John Perry and me to Dr. Ernst Gräfenberg's 1950 article, "The role of the urethra in female orgasm,"[4] in which he described a sensitive area felt through the anterior wall of the vagina, and an expulsion of fluid from the urethra that is different from urine, when this area is stimulated. In order to document that there is a sensitive area felt through the anterior vaginal wall, we had physicians and nurse practitioners do a sexological examination of the vagina looking for areas of sensitivity. They found a sensitive area that swelled when it is stimulated in the 400 women examined. These women reported that this sensitive area was felt through the anterior vaginal wall at between 11 and 1 o'clock, which was confirmed by the sexological examination. This confirmed Dr Gräfenberg's finding and led to our re-discovery of and naming this sensitive area felt through the anterior wall of the vagina, the Gräfenberg spot or G spot.[5,6]

This sensitive area is usually located about halfway between the back of the pubic bone and the cervix, along the course of the urethra. It swells when it is stimulated, although it is not possible to palpate in an un-stimulated state. This is why it is not found in a Gynecological exam. Physicians do not sexually stimulate patients, and it is also blocked by a bi-valve speculum. We hypothesized that this area is composed of many tissues, organs, and nerve pathways.[5,6]

More recently a group of researchers studied twenty women using ultrasound and reported a correlation between vaginal orgasm and the thickness of the clitoris-urethra-vaginal complex, which they reported is also known as the G spot.[7] In our latest book we stated that this highly complex region may include the anterior vaginal wall, the urethra, the Skene's glands (female prostate gland), portions of the clitoris, perhaps other glands in this region, and the surrounding muscle and connective tissue.[8] We stated that "the effect of G spot stimulation might primarily be the result of stimulation of just one structure (such as the female prostate gland) or it might be the result of stimulation of several sensitive structures that are close together." [8, p. 101]

The Gräfenberg spot has not been found universally by all researchers who have conducted sexological examinations of the vagina. It may be that the lack of universality is due to the different methods of stimulating or different criteria for identifying this area.

I had the honor a few years ago of meeting Dr. Gräfenberg's medical assistant in New York City, who was 93 at the time we met. She worked with Dr. Gräfenberg for 10 years, when he had an OB/GYN practice in New York City. My husband filmed us talking and this video will be in the Kinsey Institute at Indiana University, along with my archives. When she died last year, her daughter gave me two of the original Gräfenberg rings, the first IUDs, which he developed. See [9] for more information about the life and work of Dr. Gräfenberg.

We found that some of our subjects were women who only lost fluid at orgasm or during sexual stimulation and these women seemed to have very strong pubococcygeus muscles. So we designed a study to determine if there was a significant difference in the muscle strength of women who claimed to ejaculate and those who did not. There was a statistically significant difference in the pelvic muscle strength of women who claimed to ejaculate compared to those who did not ejaculate a fluid from their urethra, those who ejaculated had statistically significant stronger pubococcygeus muscles.[10]

The phenomenon of female ejaculation refers to expulsions of fluid from the urethra that is different from urine. Many women reported having surgery to correct this "problem" and others reported that they stopped having orgasms to prevent "wetting the bed." The fluid was described as looking like watered down fat free milk, tasting sweet, and usually about 3-5 cc's in volume.

A number of studies have been published in which the fluid expelled from the urethra has been subjected to chemical analysis. In our first published studies, we found a statistically significant difference between urine and female ejaculate in terms of prostatic acid phosphatase, urea, and creatinine.[11,12] We have also found a significant elevation in glucose in the ejaculate[11,12] and other researchers report a significant elevation in fructose.[13,14] Dr. Cabello from Spain, reported that he tested the hypothesis that all women ejaculate, although, because the amount is so small, and most women are lying on their backs during sexual activity, it may not be expelled and some may have retrograde ejaculation. He found a significant difference in Prostatic Specific Antigen (PSA) between pre-orgasmic and post-orgasmic urine specimens (PSA is what they test for in blood when screening for prostate cancer in men and measuring disease progress and remission). Cabello also found PSA in the female ejaculate.[14] Dr. Zaviacic from Slovakia has since reported on PSA being secreted by the female prostate.[13]

A study published in 2011 in the Journal of Sexual Medicine conducted in Guadalajara, Mexico by Alberto Rubio Casillas from the

Universidad de Guadalajara and Dr. Emmanuele Jannini from Italy, demonstrates that female ejaculation and squirting/gushing are two different phenomena.[15] They state that "the real female ejaculation is the release of a very scanty, thick, and whitish fluid from the female prostate, while the squirting is the expulsion of a diluted fluid from the urinary bladder."[15, pages 3502-3503] They conducted biochemical studies of the two types of fluids as well as urine from the same subject and there were significant differences.

Zaviacic proposed that the paraurethral and Skenes glands are the female prostate gland and this is where the female ejaculation is coming from.[13] The name of these glands has officially been changed to the Female Prostate Gland. We also believe that this tissue is part of the area that we have identified as the Gräfenberg spot or G spot.

Based on these findings, it is evident that some women expel a fluid from their urethra that is different from urine during sexual activities and orgasm and some women may expel a little urine. In some women G spot stimulation, orgasm, and female ejaculation are related, while in other women they are not related.[16] Some women have reported experiencing ejaculation with orgasm from clitoral stimulation and some have reported experiencing ejaculation without orgasm.[16,11] This phenomenon is reported by most women who experience it as extremely pleasurable. I hope that women who enjoy this will not have surgery designed to eliminate it and they also won't have injections (called the G shot) into the area of the G spot to supposedly enhance this sensitive area (this injection procedure has not been tested in double blind placebo controlled clinical trials, and no studies have been published, but is being offered all over in the United States and internationally).

We published our studies in 1981[10,11] and in the first edition of *The G spot and Other Discoveries about Human Sexuality* in 1982.[5] Dr. Vern Bullough, a sexuality researcher and a nurse, spoke with me at a Society for the Scientific Study of Sexuality (SSSS) meeting and said I must get my doctorate in a hard science.

In 1981 I wanted to go to the World Congress of Sexology in Israel but could not afford to go, so I wrote for copies of the published abstracts that were of interest to me. One person I communicated with was

Dr. Barry Komisaruk, from Rutgers University in New Jersey. He had heard about our research from Dr. Benjamin Graber (John Perry and I wrote two of the chapters in Graber's book, *Circumvaginal Musculature and Sexual Function*).[17,18] Dr. Komisaruk invited me to teach a class at Rutgers about our research.

Dr. Komisaruk had conducted an extensive series of studies in laboratory rats that demonstrated that vaginal mechanical stimulation produced a strong pain blocking effect, stronger than 10 mg of morphine per kg of body weight. However, the most convincing evidence that vagino-cervical stimulation blocks pain requires a verbal confirmation from women. And I wondered is the G spot just for pleasure or does it have an adaptive significance?

So this is where I went for my PhD in Psychobiology with a major in Neurophysiology (The hard science that Dr. Vern Bullough suggested).

Consequently, we performed a series of studies in women, measuring pain thresholds during vaginal self-stimulation of the area of the Grafenberg spot.

We found that the elevation in pain detection threshold increased by a mean of 47% when pressure was self-applied to the anterior vaginal wall (the area of the Gräfenberg spot). When stimulation was self-applied in a pleasurable manner, the pain threshold was greater (by 84%) than that in the resting control condition. The pain detection threshold increased by a mean of 107% when the women reported orgasm.[19] There were no increases in tactile (or touch) thresholds. This demonstrates that the effect was analgesic not an anesthetic effect and not a distracting effect.[19]

This analgesic effect was produced by pressure and by pleasurable self-stimulation applied to the anterior vaginal wall (G spot).[19,20] We then demonstrated that an analgesic effect also occurs naturally during labor.[21]

We believe that childbirth would be more painful without this natural pain blocking effect, which is activated when the pelvic, the hypogastric, and possibly the sensory vagus nerves are stimulated as the cervix dilates and from pressure in the vagina produced by the emerging fetus.

I completed my PhD in Psychobiology with a major in Neurophysiology in 1986. I obtained a second Masters degree in Nursing, because Rutgers College of Nursing wanted me on their faculty and offered me a large grant to build a human physiology laboratory, but I needed a Masters degree in Nursing. So I completed that degree the next year and started teaching and conducting further research at Rutgers University in 1987.

My research program has been devoted to validating the report of pleasurable experiences from sensual and sexual stimulation in women. Another type of orgasmic response we measured in my new human physiology laboratory was orgasm from imagery alone. That is no one, including the woman herself, touched her body, but she experienced orgasm. The physiological correlates of orgasm, that is significant increases in blood pressure, heart rate, pupil diameter, and pain thresholds, were the same during orgasm from genital self-stimulation and orgasm from imagery alone.[22]

We continued our research program by validating the subjective reports of women with complete spinal cord injury (SCI) that they do indeed experience orgasm. These women have been told, based on the literature, that they could not experience orgasm or if they did, it was a "phantom orgasm." We have documented that women with complete spinal cord injury do indeed experience orgasm from self-stimulation of the anterior wall of the vagina, the cervix, and a hypersensitive area of their body above the level of their injury.[23, 24, 25, 26, 27]

In our subjects with complete spinal cord injury, significant responses were observed in women with an injury above the level of entry of the known genital spinal nerves (pudendal, pelvic, and hypogastric). To account for this unexpected and surprising finding, we postulated the existence of a sensory pathway that bypasses the spinal cord, carrying sensory input from the vagina and cervix directly to the brain, which we postulated to be the vagus nerve.

To test whether the vagus nerve provides a vaginal sensory pathway in women, we hypothesized that the brain regions to which the vagus nerve projects (the nucleus tractus solitarius [NTS] in the medulla oblongata) would be activated by cervical self-stimulation in women with complete SCI above the level of the entry into the spinal cord of the genital sensory nerves that enter the spinal cord.

We tested the hypothesis that the vagus nerve can convey afferent activity from the cervix to the nucleus tractus solitaris (NTS) by conducting PET scans of the brain coupled with MRI to provide neuroanatomical localization.[28]

The vagus nerve projects to the NTS. During cervical self-stimulation in a woman with complete SCI at T8 using Positron Emission Tomography (PET) scan of the brain we found activation of the NTS.[28] The main problem with the PET method is low resolving power. That is, we can see that the general region of the NTS is active, but we cannot localize the region of activity restricted to the NTS. We needed an imaging method with better resolving power. We then tested and demonstrated that functional MRI (fMRI) would have adequate sensitivity as well as resolving power.

In five of the women with complete spinal cord injury in whom we recorded fMRI during cervical self-stimulation, each showed responses in the same NTS region of the medulla. These findings show that the NTS, which is the sensory nucleus of the vagus nerve, responds to cervical self-stimulation even if the genital sensory pathways through the spinal cord are completely severed.[29,30]

The NTS is just one of the brain regions activated by cervical self-stimulation. Other brain areas that were activated in women with and without complete spinal cord injury were the cingulate cortex, nucleus accumbens, amygdala, basal genglia, insula, hippocampus, and the paraventricular nucleus of the hypothalamus.[30,31]

There was a gradual increase in activity in specific brain regions leading up to orgasm. The first 2-minute period of cervical self-stimulation only the amygdala, basal ganglia, and insula were activated, as orgasm started the cingulate cortex and nucleus accumbens were also activated, and all seven regions were activated at orgasm (see [31] for review).

We also conducted fMRI's of the brain in able-bodied woman during an orgasm induced by imagery alone in the absence of genital physical stimulation and also by genital self-stimulation, in a counter-balanced design. The NTS as well as the same areas of the brain identified above were activated.[29]

We are planning to use fMRI images as a biofeedback to the women to determine if they can see their own specific regions of brain activated, could they voluntarily intensify or decrease the activity. This may be helpful in control of pain and in women with Persistent Genital Arousal Disorder (PGAD). This "neurobiofeedback" system could have interesting therapeutic applications, if it turns out that we are able to control our brain activity by observing it directly.

A review of these studies can be found in our book, *The Science of Orgasm.*[31]

These findings suggest that orgasm in women is in the brain, it is felt in many body regions, and it can be stimulated from many body regions as well as from imagery alone. Orgasm is not a just a reflex, it is a total body experience. We need to continue to be open to documenting the various sensual and sexual experiences reported by women.

People need to be encouraged to feel good about the variety of ways they may experience sexual pleasure, without setting up specific goals, such as finding the G spot or experiencing female ejaculation. Healthy sexuality begins with acceptance of the self, in addition to an emphasis on the process, rather than only the goals, of sexual interactions.

We are continuing our research to look at fMRI's of the brain in men and in women. We plan to study women before and after hysterectomy and men before and after prostate surgery.

We continue conducting research concerning the pleasurable responses of women to different forms of sensual and sexual experiences to validate women's subjective reports. Most of the research that has been conducted has been conducted in men and the findings extrapolated to women. Women are different from men and my whole research program has been devoted to documenting and validating the experiences that women report are pleasurable. There is not one sensual and sexual response pattern in women, and women are capable of experiencing many sensual and sexual responses.

Sexual Health Organizations

I was very fortunate in that the professional organizations not only accepted my proposals to speak about our research but my colleagues gave me so much encouragement and feedback. I felt I had to give back to the organizations for all of their support. I did this through my volunteer work on their boards. I served on the AASECT Board of Directors for over 10 years and was President of AASECT from 1998-2000. I served on the Society for the Scientific Study of Sexuality (SSSS) Board of Directors for 8 years, and was president of SSSS from 2002-2003. Currently, I am the only person who has been president of both organizations. I have been on the Board of Directors of the Foundation for the Scientific Study of Sexuality (FSSS) for many years and they have given me an emeritus status. I was on the Executive Committee of the World Association for Sexual Health (WAS) for 12 years, I was Vice president from 2001 to 2005 and Secretary General/Treasurer from 2005 to 2009. I declined serving as President of WAS.

The Future

Where do I go from here? I hope to spend more time with our family. My husband Jim and I have our bucket lists, and we are being more selective about where and how we spend our time. I have been so blessed, in that I have been to 97 countries, and have been invited to speak about my research in most of the places we have visited. I hope this short overview of my life as a sexuality educator and researcher will help others to continue learning more about men and women and their pleasurable sensual and sexual experiences.

References

1. Whipple, B., & Gick, R. (1980). A holistic view of sexuality: Education for the professional. *Topics in Clinical Nursing*, *1*, 9198.
2. Kegel, Arnold H. (1949) "The Physiologic Treatment of Poor Tone and Function of the Genital Muscles and of Urinary Stress Incontinence," *Western Journal of Surgery, Obstetrics, and Gynecology*, *57*: 527-535.

3. Kegel, Arnold H. (1952) "Sexual Functions of the Pubococcygeus Muscle," *Western Journal of Surgery, Obstetrics, and Gynecology, 60*: 521-524.
4. Gräfenberg, Ernst. (1950) "The Role of the Urethra in Female Orgasm" *International Journal of Sexology 3*: 145-148.
5. Ladas, A.K, Whipple, B., & Perry, J.D. (1982) *The G spot and Other Recent Discoveries about Human Sexuality*, New York: Holt, Rinehart and Winston (published in 22 languages).
6. Ladas, A.K, Whipple, B., & Perry, J.D. (2005) *The G spot and Other Discoveries about Human Sexuality*, Classic Edition, New York: Owl Books.
7. Gravina, G.L., Brandetti, F., Martini, Pl, Caros, E., Distasi, S.M., Morano, S., Lenzi, A., & Jannini, E.A. (2008) "Measurement of the thickness of the urethrovaginal space in women with or without vaginal orgasm" *Journal of Sexual Medicine 5*: 610-618.
8. Komisaruk, B.R., Whipple, B., Nasserzadeh, S., & Beyer-Flores, C. (2010) *The Orgasm Answer Guide* Baltimore, Maryland: Johns Hopkins University Press.
9. Whipple, B. (2000). "Ernst Gräfenberg: From Berlin to New York" *Scandinavian Journal of Sexology, 3* (2), 43-49.
10. Perry, J.D., & Whipple, B. (1981). Pelvic muscle strength of female ejaculators: Evidence in support of a new theory of orgasm. *The Journal of Sex Research, 17*, 2239.
11. Addiego, F., Belzer, E.G., Comolli, J., Moger, W., Perry, J.D., & Whipple, B. (1981). Female ejaculation: A case study. *The Journal of Sex Research, 17*, 1321.
12. Belzer, E., Whipple, B., & Moger, W. (1984). On female ejaculation. *The Journal of Sex Research, 20*, 403406.
13. Zaviacic, M. 1999 *The Human Female Prostate: From Vestigial Skene's Paraurethral Glands and Ducts to Woman's Functional Prostate.* Bratislava, Slovakia: Slovak Academic Press.
14. Cabello, F (1977) Female ejaculation: Myths and reality, pp. 1-8. In *Sexuality and Human Rights.* Borras-Valls JJ, Perez-Conchillo, M. (Eds.) Valencia, Spain. Nau Llibres.
15. Rubio-Casillas, A., & Jannini, E.A. (2011). New Insights from One Case of Female Ejaculation. *Journal of Sexual Medicine, 11*: 8: 3500-3504.
16. Whipple, B., & Komisaruk, B.R. (1991). The G spot, orgasm, and female ejaculation: Are they related? In P. Kothari (Ed) *The Proceedings of the First International Conference on Orgasm.* (pp. 227237) Bombay, India; VRP Publishers.
17. Perry, J.D., & Whipple, B. (1982). Multiple components of the female orgasm. In B. Graber (Ed.) *Circumvaginal Musculature and Sexual Function* (pp. 101 114). New York: S. Karger.
18. Perry, J.D., & Whipple, B. (1982). Vaginal myography. In B. Graber (Ed.) *Circumvaginal Musculature and Sexual Function* (pp. 6173). New York: S. Karger.

19. Whipple, B., & Komisaruk, B.R. (1985). Elevation of pain thresholds by vaginal stimulation in women. *Pain, 21*, 357367.
20. Whipple, B., & Komisaruk, B.R. (1988). Analgesia produced in women by genital selfstimulation. *The Journal of Sex Research, 24*, 130140.
21. Whipple, B., Josimovich, J.B., & Komisaruk, B.R. (1990). Sensory thresholds during the antepartum, intrapartum, and postpartum periods. *International Journal of Nursing Studies. 27*, (3), 213221.
22. Whipple, B., Ogden, G., & Komisaruk, B.R. (1992). Physiological correlates of imagery induced orgasm in women. *Archives of Sexual Behavior, 21*(2), 121 133.
23. Whipple, B., Gerdes, C.A., & Komisaruk, B.R. (1996). Sexual response to self-stimulation in women with complete spinal cord injury. *Journal of Sex Research,33*(3). 231-240.
24. Komisaruk, B.R., & Whipple, B. (2005). Brain activity imaging during sexual response in women with spinal cord injury. In J. Hyde (Ed.) *Biological Substrates of Human Sexuality*. (pp. 109-145) Washington DC: American Psychological Association. 26
25. Whipple, B., Richards, E., Tepper, M., & Komisaruk, B.R. (1996). Sexual response in women with complete spinal cord injury. In D.M. Krotoski, M. Nosek, & M. Turk (Eds.) *Women with Physical Disabilities: Achieving and Maintaining Health and Well-Being* (pp. 69-80). Baltimore: Paul H. Brooks Publishing Co.
26. Whipple, B., Richards, E., Tepper, M., & Komisaruk, B.R. (1966). Sexual response in women with complete spinal cord injury. *Sexuality and Disability, 14*(3), 191-201.
27. Whipple, B., Richards, E., Tepper, M., Gerdes, C., & Komisaruk, B.R. (1998). A quantitative and qualitative study concerning sexual response in women with complete spinal cord injury. In. J.J. Borras-Valls & M. Perez-Conchillo (Eds). *Sexuality and Human Rights* (pp. 267-271). Valencia, Spain: World Congress of Sexology.
28. Whipple, B., & Komisaruk, B.R. (2002). Brain (PET) responses to vaginal-cervical self-stimulation in women with complete spinal cord injury: Preliminary findings. *Journal of Sex and Marital Therapy, 28*: 79-86.
29. Komisaruk, B.R., Mosier, K.M., Liu, W-C, Criminale, C., Zaborszky, L., Whipple, B., & Kalnin, A. (2002). Functional localization of brainstem and cervical spinal cord nuclei in humans with fMRI. *American Journal of Neuroradiology, 23*: 609-617.
30. Komisaruk, B.R, Whipple, B., Crawford, A., Grimes, S., Liu, W-C., Kalnin, A., & Mosier, K. (2004) Brain activation during vaginocervical self-stimulation and orgasm in women with complete spinal cord injury: fMRI evidence of mediation by the Vagus nerves. *Brain Research, 1024,* 77-88.
31. Komisaruk, B.R., Beyer-Flores, C., & Whipple, B. (2006). *The Science of Orgasm.* Baltimore, MD. Johns Hopkins Press.

Changing Landscapes at the Top of the World

Ulyana N. Horodyskyj

Ulyana Nadia Horodyskyj currently is a Research Associate at the Cooperative Institute for Research in Environmental Sciences (CIRES) in Boulder, CO. She graduated with her PhD in geological sciences from the University of Colorado Boulder in May 2015. Since 2011, she has led expeditions in the Nepalese Himalaya, attempting to quantify glacial lake growth and evolution, as well as, impacts of pollution to snow and ice melt at high altitudes. Ulyana has traveled and worked on all 7 continents, most memorably working as a 'deck rat' on an icebreaker in Antarctica when she was 21 years old. In her free time, she enjoys writing, cooking, climbing and mountain biking.

It started as a dull roar in the distance. Snow and ice, crashing down a slope at 3000 meters in the Swiss Alps, captured my attention as it got louder and louder. I stopped throwing snowballs at my mother long enough to stare in awe. I was 6 years old and had just witnessed my first avalanche. I had no idea that such early exposure to the mountains, glaciers, and their hazards would influence my life, and finally come full circle 22 years later on a science research expedition to Mt. Everest.

My journey of scientific discovery began at quite a young age. In my pre-teens, I started dabbling with various science projects. From collecting and weighing junk mail, to studying the effects of solar radiation on t-shirt materials, to calculating cereal volumes compared to cost, my childhood was filled with exploration. I loved asking questions and searching for the answers through creative means. In the case of the junk mail project, I discovered Tuesdays always had the most (by weight) and by the end of the year, the pile of mail weighed more than me! T-shirts do not offer as much UV protection as you would expect/hope. And, I found that you are not getting your money's worth with most cereals geared towards kids, such as Kix, while the 'healthy' cereals, like Raisin Bran and Wheaties, did the best. Mom was glad to know.

During my high school years, with the help of my father and project mentor, the late astrophysicist Dr. Robert L. Forward, I explored solar sail technology – a way to travel in space without needing fuel. By creating my own computer simulations, I discovered how quickly I could get to Mars using a certain payload and solar sail area (e.g., sometimes, it could be no smaller than the size of an American football field). That project work launched me into a world of symposia and science fairs across the nation throughout my teens, earning me enough scholarship money to attend my dream school: Rice University. But more importantly, it allowed me to learn about true discovery – about diving into the scientific method, while using curiosity to guide me, and inventing tools to solve problems. This would serve me well as I continued to pursue science at a higher level.

While completing a Masters degree in planetary geology at Brown, I met David Breashears, director of the Everest IMAX film and founder of a non-profit called GlacierWorks, which uses high-resolution photography to document the changing landscape of the Himalaya. When I was 12 years old, I saw the film in the IMAX Theater with my middle school class. I still remember the footage of a climber crossing a crevasse, which I found both frightening and exciting. I never dreamed I would meet, let alone work alongside, the director of the movie. With his encouragement, I applied to the University of Colorado Boulder, where Dr. Roger Bilham, the geology 'star' in David's Everest and Kilimanjaro IMAX films would become my PhD advisor. At this point in my life, I knew I wanted adventure and science as a career path – and what better way to pursue that than in the Himalaya with two great mentors?

Figuring out a PhD research project was no easy task, though. I came to Colorado with big hopes for fieldwork and research, but initially no money to do so. By the end of it, my PhD was paid for by an unconventional combination of small grants from the university, USAID, the Geological Society of America, The Explorers Club, and generous donations from project sponsors through crowd fundraising. The latter contributions allowed me to share with the public the joys and frustrations of my field expeditions, and my discoveries along the way. In the spirit of my first mentor, I was 'paying it forward', in the form of scientific knowledge both to the lay public and to the locals I trained while living and working in Nepal.

In late 2010, while perusing David's high-resolution photography from Nepal and Tibet, I noticed that when comparing earlier imagery to the newer imagery in the 2000s, there were a lot more lakes dotted across the surfaces of the glaciers. Not only that, the vertical height of the glacier had dropped tens of meters. I was amazed at how much ice mass loss I saw, in such a short amount of time. Thinking perhaps these lakes had something to do with it, I started searching through the scientific literature. I learned that these were called 'supraglacial' lakes, since they were perched on the surface of the glacier, with ice still underneath. They were 'dirty' lakes, with lots of mud, and the whole lower tongue of the glacier (called the ablation zone) was in fact ice covered in heavy rocky debris, sometimes meters thick. This is quite different from the 'typical' clean ice glaciers you find in Greenland and Antarctica. Debris, if thick enough, acts as an insulating blanket and prevents the ice from melting. But, if too thin (< 2 cm), the debris absorbs more solar radiation and enhances melting of the ice below (e.g., Ostrem, 1959; Nakawo and Young, 1982; Nicholson and Benn, 2006).

After months of research, I decided to focus on the Ngozumpa glacier, one of Nepal's largest and longest (at 18 kilometers), which had seen initial studies on its debris cover, supraglacial ponds and its large 'Spillway' terminal lake by teams from the UK (e.g., Benn et al., 2000; Benn et al., 2001; Nicholson and Benn, 2006; Thompson et al., 2012). I headed out to the Himalaya in June/July 2011 to study these smaller ponds with a whitewater raft, a handful of time-lapse cameras to take photos on an hourly basis, and various instruments Roger and I had assembled in the lab. The ~50 kilometer journey up to over 5000 meters with my field assistant, Ang Phula Sherpa, was grueling. My pack was not heavy, but it certainly felt like it was gaining weight as we climbed higher and higher along the dusty trails. It was quite humbling to be passed by yaks. About a week later, we finally reached our base camp, the Gokyo village, where we would stay with a Sherpa family for the next month while my body got used to the thin air.

Even if I could not keep pace with Phula or catch my breath easily, I was excited to begin exploring the region. Prayer flags fluttering in the wind, yak bells jingling in the early morning hours, and a fog creeping up the valley in the afternoon hours added to the mysticism of this Himalayan landscape. Catching my first glimpse of the glacier, I was surprised at how extensive and thick the debris cover was – the only visible ice was on the walls surrounding about half a dozen supraglacial ponds in our immediate vicinity. While down on the surface, we quickly learned how to spot glacial 'quicksand' along the shorelines of the ponds. Doing anything at altitude is challenging, let alone tugging a 20-kilogram raft out of the mud.

To view a map of The Ngozumpa glacier and surroundings see: Taking the pulse of Ngozumpa, which discusses my research there - www.bbc.com/news/science-environment-16317090

One day, my stomach was not feeling too well, so we did not make it down to the glacier until around lunchtime. We soon found out, though, that despite the delay, luck was on our side. Earlier that morning, some porters who were crossing the glacier to another village reported hearing a very loud collapse. Morning fog prevented them from seeing where. As Phula and I made our way down there, we saw that one of the ponds (called 'Ucam') had lost 20 vertical meters of ice from one of its surrounding ice walls. There were chunks of ice the size of small cars in the water and we found our raft 10 meters further on-shore, flipped over, but thankfully still intact. One of my waterproof Pelican cases was found in the middle of the pond. The instrument itself had been tethered to the bottom, while its electronics box had floated away. A rescue mission was quickly mounted and we found that not only had the instrument – a pressure transducer – survived, but also recorded the event as an 8 centimeter rise in the lake. Though that may not sound like a large number, that is a rise in water level that may be expected after a few weeks of misty rain at this elevation – and one ice calving event had caused it. I was humbled by the power of nature and thankful that we were not out in the boat earlier, when the collapsing ice might have struck us.

In another pond ('Dcam') further up the glacier, we were in the raft deploying some instruments when we spotted a large boulder starting to roll off the top of a 40-meter ice cliff. No words were exchanged – we just started paddling hard. Thankfully, the boulder missed us and the waves generated from the collapse were not strong enough to flip us over. Keeping a wary eye on the ice walls, we paddled to shore to take some high-resolution photographs of that ice wall with its interesting 'bathtub ring' appearance, as well as to collect samples of glacial 'flour' (very fine-grained mud) that was left behind on its shore. A few days later, during particularly bad weather, Phula and I ventured out to download time-lapse photos from an overhead camera, since we could not get down to the glacier surface safely. We could not even see the pond below! After analyzing the imagery, which had been shot on an hourly basis, I found that the pond had initially filled 10 meters, drained in two separate events, leaving behind the 'bathtub ring' marks and flour, and then finally refilled to almost the same height – all in the span of a week. Given satellite image frequency (every 2 weeks) and heavy cloud cover during the monsoon in this region, we would not have seen this event had it not been for the oblique camera view. The significance of this event is that the equivalent of ~40 Olympic-sized swimming pools disappeared in a matter of days somewhere down the glacial system and similar drainage events continued throughout the summer: (http://www.bbc.com/news/science-environment-16317090).

Understanding how these ponds behave is important when trying to forecast future flooding hazards. In just a short few years during the course of my PhD research, these small ponds showed horizontal expansion rates in the tens of meters and Dcam pond was the deepest in the region, at 54 meters (Horodyskyj, 2015). Someday, they may coalesce into a 'Spillway' situation, as we already see with the large lake near the terminus of Ngozumpa glacier, which started growing in 2001 (Thompson et. al., 2012). Other glaciers in the Everest region are growing even larger lakes, such as Imja and Tsho Rolpa, which hold millions of cubic meters of water and pose a significant glacial lake outburst flood threat (e.g., Chikita et al., 1999, 2001; Watanabe et al., 1995). Spillway Lake fortunately is still in the adolescent phase of growth; hence, its study is timely. There is still time to mitigate its hazard.

A Fulbright fellowship allowed me to live abroad in Nepal from August 2013 – June 2014. I was able to visit my pond field sites more frequently than in prior years, as well as begin a long-term study on the Spillway Lake. Frequent visits to this area are important, given the effort needed to install equipment under harsh conditions. These conditions can adversely affect the electronics of the instruments and limit their service life. During the summer 2013, I had a weather station and camera tracking the icefalls of Cho Oyu (the source of Ngozumpa glacier) fail at 5200 meters. You can imagine the heartbreak at discovering the data were irretrievable. The silver lining is that instead of losing a whole year of data, only a few months were lost.

At Spillway Lake (4800 meters), in addition to time-lapse photography, my field assistants and I installed a long-term weather station, deployed temperature and water level sensors (with downloads every few months), and used a sonar system in open water to create 3D maps of the lake sub-basin floors. The latter effort was the biggest team I managed – 8 Westerners and Sherpas – and was completed in June 2014, when the lake had thawed but before the start of the monsoonal rains. During off-season times in the valley, sometimes it was just me out on the glacier, downloading data, taking new photos, or inevitably fixing something, given the very low air temperatures and dry, dusty conditions. The worst experiences out there included getting caught in a fierce glacial dust storm, where I could not easily see or breathe; and nearly drowning at the outlet of the glacier, when I flipped a kayak and was submerged in 2 °C water.

After analyzing a year's worth of data, I found that while some areas of Spillway Lake had the potential to deepen significantly (a few meters per year), given warmer water temperatures and thin mud layers on the bottom (or, possibly bare ice), other areas were found to be stable, given cold water temperatures (e.g., < 1 °C) with overlying layers of rock and mud. The lake floor differences were found to be quite heterogeneous across Spillway Lake's sub-basins, meaning that forecasting change is not straight-forward.

The so-called 'Main' sub-basin of Spillway has been showing the most rapid growth since the 2000s and is an area that warrants further study (e.g., Thompson et al., 2012). In November 2013, I found an odd feature on the surface, while climbing up a 'debris island' to check on the weather station's battery level. While most of the lake surface was already frozen, there was one spot with a perfectly circular unfrozen hole. Surrounding this natural-looking 'fishing hole', there were concave 'bubbles' frozen into the ice from below. Upon return to the site in December 2013, I found that the hole had frozen over, but its ice was thinner than the surrounding ice. This feature is interpreted to be evidence of an 'upwelling,' which occurs due to temperature and density differences of the source water. Fresh meltwater emerging from an ice tunnel in the depths of the glacier, while colder, is cleaner and less dense than the muddy lake water, thus causing the meltwater to 'upwell' to the surface given the density difference. Similar upwelling spots, but smaller in scale, were found throughout the lake during the course of the year. They may be critical for predicting future rapid lake deepening, as likely there is minimal debris (rock and mud) overlying the ice in those areas, allowing the warmer water to melt the ice faster.

During my Fulbright year, I also was able to launch a new phase of my project, which was aimed at understanding the impacts of contaminants on snow and ice melt. While ponds and lakes are growing in the ablation zones of glaciers, how are their accumulation zones, where snow turns to ice, faring? Pollution from black carbon (more commonly called 'soot') and dust are composed of dark particles and when on a lighter surface such as snow, they can absorb more solar radiation, thus leading to enhanced melting of that surface (e.g., Warren and Wiscombe, 1985; Hansen and Nazarenko, 2004). This is similar to the effect of rock/mud debris on glacier ablation zones, as noted earlier. Higher melting rates can affect a glacier's ability to retain snow (to turn into ice) within its accumulation zone, as well as create climbing hazards, by turning the snow into slush. Sources for black carbon include industry (factories), road construction, cars, motorbikes, forest fires, and other land use change, while dust can be naturally occurring or a derivative of land use change (e.g., Goldberg et al. 1985).

I was able to collect and filter snow samples from nearby lower peaks, mountain passes, and villages ranging from 4500 – 6000 meters in all seasons for both eastern and central Nepal, though the conditions could be quite challenging – bitter cold temperatures, bouts with altitude sickness, blizzards, and avalanches. In more remote regions, I found dust was the dominant contaminant, while closer to the villages, black carbon, or, 'soot', could still affect the peaks, even at over 6000 meters in elevation (Horodyskyj, 2015). In the off-seasons of climbing (December – February), contaminants were found in large quantity in the snowpack due to the presence of a regional 'Asian brown cloud,' a layer of air pollution, from combustion and biomass burning. These contaminants would sit in the lower valleys as there's no rain during this time to help wash out the contaminants (e.g., Barros and Lang, 2003). The next phase of this work was to go to 8000 meters, to see what kind of effect these small particles might have in the upper reaches of glacier accumulation zones.

In the early morning hours of April 18, 2014, I rolled over in bed in Dingboche, a small 4500-meter Himalayan village in Nepal. It was a restless night. Our team from the American Climber Science Program (ACSP), a nonprofit organization that brings together scientists, climbers, and other volunteers to work in the high alpine environment, was finally trekking to Everest base camp to set up for project work higher on the mountain. This was to be the culmination of my year long initiative towards understanding the impacts of contaminants on snow melt – and at my highest altitude yet.

Sadly, an icefall avalanche took the lives of 16 Nepali climbers earlier that day, including Asman Tamang, a young Nepali member of our ACSP team. I remember the news trickling down the valley to us. I remembered thinking avalanches are quite common here, but did not realize the magnitude of the tragedy until we arrived at our camp and saw a helicopter long-lining bodies down from the mountain. Another very striking image that stayed with me was that of valley 'smog' seen from Everest base camp in the early morning hours following the tragedy. Even at these high, supposedly 'pristine' elevations, the impact of human activity was clear.

The following year, I briefly contemplated returning to Everest in the spring to complete my project work. But, I did not feel right going to the mountain so soon after a tragedy of that magnitude and decided to focus my efforts on finishing my dissertation. In April 2015, a 7.8 magnitude earthquake rocked the Himalayan nation, killing over 9,000 people. I was devastated to see a country and people I love suffering such loss. From Kathmandu to the tiny mountain villages of the Everest region, everyone I came to know during my time abroad was affected in some way. In the midst of grief, I found the only way forward was to finish my writings and make the work available and hopefully useful to the villagers. During the course of my Fulbright year abroad, I had dedicated time to building a 'Sherpa-Scientist Initiative,' to educate the locals on the changing landscape of the Himalaya, and what they can do to continue to monitor and mitigate the natural hazards in their backyards. My project work and personal investment in the people of Nepal has not ended with the PhD. In fact, I believe that my work has only begun, as there are many new scientific discoveries to be made in the region. The glacial lakes I have studied show no signs of slowing their growth. New ones may appear in the coming years. We have the chance now to be creative problem-solvers and put into action geo-engineering projects in the form of dams and siphons to mitigate the flood risks. Switching to clean burning stoves in the villages, as well as using carbon filters on cars and motorbikes in nearby cities (Kathmandu, Delhi) can help clear the air of pollutants, good for the health of the people and mountains alike.

In some ways, the landscape is as resilient as the Nepalese people have been in the face of such hardship with the recent earthquake. But it is important that we ***all*** – locals, climbers, trekkers, researchers – become stewards of our environment, and ensure the natural beauty that is the Himalaya stays that way for generations to come.

For more about the project work, please see:
http://howonearthradio.org/archives/tag/ulyana-horodyskyj
http://www.cpr.org/news/story/remote-mountain-peaks-are-boulder-scientists-laboratory
http://www.scientificamerican.com/author/ulyana-horodyskyj/
http://blackicehimalaya.wordpress.com/

References

Barros, A. P. and T.J. Lang, (2003), Monitoring the Monsoon in the Himalayas: Observations in Central Nepal, June 2001, *Mon. Weather Rev*. 131, 1408–1427, doi:10.1175/1520-0493.
Benn, D.I., Wiseman, S., and C.R. Warren, (2000), Rapid growth of a supraglacial lake, Ngozumpa Glacier, Khumbu Himal, Nepal, *International Association of Hydrological Sciences Publication* 264 (Symposium at Seattle 2000 – Debris-Covered Glaciers), 177-185.
Benn, D.I., Wiseman, S., and K.A. Hands, (2001), Growth and drainage of supraglacial lakes on the debris-mantled Ngozumpa Glacier, Khumbu Himal, Nepal, *Jour. Glac.,* 47, 626–638.
Chikita, K.A., Jha, J., and T. Yamada, (1999), Hydrodynamics of a supraglacial lake and its effect on the basin expansion: Tsho Rolpa, Rolwaling Valley, Nepal Himalaya. *Arctic, Antarctic, and Alpine Research,* 31, 58–70.
Chikita, K.A., Jha, J., and T. Yamada, (2001), Sedimentary effects on the expansion of a Himalayan supraglacial lake. *Global and Planetary Change,* 28, 23–34.
Goldberg, E., (1985), Black carbon in the environment, Wiley and Sons, New York, NY.
Hansen. J. and L. Nazarenko, (2004), Soot climate forcing via snow and ice albedos, *Proceed. Natl. Aca. Sci.,* 101(2), 423-428.
Horodyskyj, U.N., (2015), Contributing Factors to Ice Mass Loss on Himalayan Debris-Covered Glaciers, *Ph.D. dissertation*, ProQuest, UMI Dissertations Publishing, 183 pg.
Nakawo, M., and G.J. Young, (1982), Estimate of Glacier Ablation Under a Debris Layer from Surface Temperature and Meteorological Variables, *Jour. Glacio.,* 28(98), 29-34.
Nicholson, L.A., and D.I. Benn, (2006), Calculating ice melt beneath a debris layer using meteorological data. *Jour. Glac*., 52 (178), 463–470.
Ostrem, G., (1959), Ice Melting under a Thin Layer of Moraine, and the Existence of Ice Cores in Moraine Ridges, *Geograf. Annal.*, 41(4), 228-230.
Ritter, M., (2010), The Physical Environment: An Introduction to Physical Geography, http://www.earthonlinemedia.com/ebooks/tpe_3e/title_page.html.
Thompson, S., Benn, D.I., Dennis, K., and A. Luckman, (2012), A rapidly growing moraine dammed glacial lake on Ngozumpa Glacier, Nepal. *Geomorphology,* 75, 266–280.
Warren, S.G. and W.J. Wiscombe, (1985), Dirty snow after nuclear war, *Nature*, 313(7), 467-470.
Watanabe, T., Kameyama, S., and T. Sato, (1995), Imja glacier dead-ice melt rates and changes in a supraglacial lake, 1989–1994, Khumbu Himal, Nepal: danger of lake drainage. *Mountain Research and Development*, 15, 293–300.

A Metamorphosis

Jeffrey A. Lockwood

Jeffrey A. Lockwood is a professor of Natural Sciences & Humanities in the Department of Philosophy and MFA Program in Creative Writing at the University of Wyoming.

I was about 10, and I was flipping through a *Time-Life* history book. There was a photograph of street urchins who'd climbed an outfield fence to watch a big league game. I think it was in New York, but it doesn't matter.

What matters is that one kid looked directly into the lens. It matters that the year was 1908 (I somehow remember this detail). And it matters that the kid would've been at least 80 by the time I saw the image. And to me, 80 was old, essentially dead.

I realized that I was going to grow old and die. That kid had his picture in a book while the others had grown up and died anonymously. At that moment, I knew that I didn't want to die without having mattered.

* * *

Things were easier in the 60s. Sure, there were civil rights battles, the Vietnam War, and the possibility of a nuclear holocaust. But, at least, I understood that boys studied science. The only real question was what kind of scientist to become.

My dad was a physicist (my mom, an artist-turned-homemaker), but I really liked animals. So I planned to become a veterinarian (not a real scientist, but close enough). I worked at a veterinary clinic through high school and became disillusioned. Insects, however, were darkly enchanting—it was possible to discover something truly original about these creatures, given that we haven't even named most of them.

And entomology was pretty funky for being a science. Not quite like being an artist. Now I teach non-fiction writing in a Master of Fine Arts program. It's not quite like being a poet, but look out, Gary Snyder.

* * *

My teachers have been:

A physicist who loved science but his son even more;
An animal behaviorist who had nearly become a monk;
A Unitarian minister who signed his letters, *Wishing you Peace & Unrest*;
An entomologist who shared his poetry and out-of-body dreams with me;
A botanist who, if not for science, would've been a musician;
A student who battled schizophrenia and won heroically, then lost tragically, then won courageously;
A department head who derived genuine, vicarious pleasure from the success of others;
A dean who couldn't make hard decisions or keep his word (not all lessons involve virtue);
A colleague who struggled with alcoholism and is winning, for now;
A poet who told me, honestly and compassionately, to stick with writing essays;
A philosopher who knew more about science than 99% of scientists.

* * *

It's true—the fights in academia are so intense because the stakes are so low. I like to ask graduate students during their defenses, what would happen if everything you said was wrong? Most of the time, a few dozen scholars would be miffed. Of course, by the time anyone figured it out (most science is most assuredly not repeated), the student would be gainfully employed or even tenured.

I think that this is why I gravitated to agriculture and pest management. The sophisticates scorned applied research—the less connected to the world of money and dirt, the better. But I relished the possibility that my work would matter—that if I screwed up, then real people would be really upset. There's not much satisfaction in performing CPR on a dummy.

* * *

I learned a great deal from numbers. I loved their precision. Even when they were inconsistent, statistics provided clarity. I conducted experiments with dozens of chemical and biological control agents, and the question was always: "How much control did we get?" When there were 22.4 grasshoppers per square yard before treatment and 4.6 after treatment, there was something wonderfully unambiguous about 79.5% control. This meant that on a standard program of 10,000 acres, there were 861,520,000 fewer grasshoppers, although I very rarely saw dead ones, which made the number seem clean and objective. Until I began to wonder what that many lives meant.

I could answer "What does a grasshopper weigh?" or "How long does a grasshopper live?" but I couldn't resist asking a question that science couldn't address: "What does a living creature mean?"

In Australia, I worked with extraordinarily aggressive gryllacridids (like grasshopper-cricket hybrids). These creatures hid in their silken nests until flushed into the open, at which point they attacked with unmitigated rage. The ones I kept in the laboratory never diminished their ferocity; they could not understand that I meant no harm. One day, I accidently trapped one between the lid and edge of the cage and ruptured its abdomen. I could not comprehend what came next—the insect began to calmly consume its own entrails.

Not long after I moved to Wyoming, a beloved graduate school mentor was killed in the course of a robbery by a young, angry man. My teacher had warned me about the hazards of ascribing my own subjective experience to that of other beings. Nonetheless, I've tried to comprehend the violence. I am still bewildered.

Audio: WNYC Radiolab Podcast with Jeff Lockwood - Killer Empathy - www.radiolab.org/story/185551-killer-empathy

* * *

I've developed courses in "Natural Resource Ethics," "The Biodiversity Crisis," "Deep Ecology," and "Environmental Justice." I've engaged nearly four thousand students in such academic ventures. But I don't know if any student is leading a richer life, whether her work is more meaningful or his decisions more responsible for my efforts.

When my sister began college, she was the sort of poorly motivated, unfocused student who frustrates me in the classroom. Not until she changed her major to nursing did her formidable potential and energy become realized. A few years ago, she was visiting a museum when a woman collapsed with a heart attack. My sister administered CPR until the ambulance arrived.

I envied the certain knowledge that one has changed a single life.

* * *

My students dutifully calculated the number of heartbeats per year and blinks in a lifetime. But the numbers weren't real to them. So, one morning I pulled out a box of graph paper with a fine grid—1/25th of an inch—and taped together a quilt of twenty-four sheets (the final one being precisely trimmed to 22,500 squares). "That's exactly a million," I declared to the class that afternoon.

The students drew close enough to focus on a single square and then stepped back. "How many sheets would it take for the human population, if each square was a person?" one asked. "It would blanket a football field," I replied after a minute of extrapolation.

They grew quiet, each envisioning a tiny square somewhere near the 50-yard line.

* * *

For a college class, my son wrote a 'spiritual autobiography' in which he recalled from his childhood: "It was, and is, easy for me to become absorbed in the grandeur and desolation of the natural world, but with gentle prodding while driving back home after a day in the field my father would ask me to expand on the emotions and experiences of having counted the dying grasshoppers in a square foot of prairie to determine the effectiveness of a particular pesticide. He would allow me to ramble as I sorted through my thoughts out loud before pushing my beliefs to their logical conclusions… Much to my consternation, I never could get away with simple responses or platitudes." I like to believe that I so affected my students. But I can rest easy, knowing that I mattered to him.

* * *

My claim to fame is having pioneered a method for controlling rangeland grasshoppers that has saved agriculture millions of dollars and decreased insecticide use by thousands of tons across the western United States. By applying low doses of insecticides to swaths interspersed with untreated areas, ranchers reduce pest numbers while preserving beneficial species and leaving enough grasshoppers behind to sustain the native predators and parasites. Ok, it's not rocket science, but we blanketed the prairie with high doses of chemicals for decades before I came along.

It was an adrenaline rush to see the method adopted across millions of acres. I figured that if it worked, I'd be a hero. It did. I wasn't. No parades or medals. Not even a hearty "thanks!" from the National Cattlemen's Association. I called the method RAATs (Reduced Agent-Area Treatments) which caught on. I should've named it JEFF (Just Enough For Functionality).

* * *

When I was a kid, I learned judo and even became reasonably good at this sport. I wasn't particularly powerful or quick, but I was pretty smart. I understood that to throw one's opponent required finding the tipping point—the place and moment at which one could, with minimal effort, exploit the other person's imbalance and momentum. Likewise, rock climbers seek 'affordances'—those finger-and-toe holds that the stone provides, rather than the brute force of hammering in hardware. Knowledge is (soft) power.

Understanding the movement of grasshoppers, the headlong rush of their populations, and what their natural enemies provide when we don't force our chemical will on the prairie allowed me to see how less can be more in developing RAATs—less poison and damage, more birds predating, bees pollinating, grass producing and, yes, ranchers profiting.

* * *

I have friends with alcoholism, depression, schizophrenia, and dissociative identity disorder. I know because they told me so. And I know that most of the people in their lives don't know. Who wants people to think you're nuts?

I've struggled with anxiety attacks periodically, and then there was that debilitating panic attack. I encountered a phenomenal grasshopper infestation in a deep draw on the rangeland and the creatures pelted me and clung to my hair and writhed into my clothing. It felt like I was smothering, just like in those childhood nightmares where I'd wake up twisted in the sheets after an amorphous presence slowly swelled to fill the room.

It's not a good thing for an entomologist to have a panic attack amidst the creatures he studies. I didn't tell anyone for years. I was afraid they'd think I was crazy.

* * *

“Close your eyes,” I asked. Two hundred lids quivered suspiciously. “Picture in your mind a football field." The exercise was safe and simple; their eyes shut.

“Now, cover the field in a tropical forest—imagine the dappled sun and the damp shade.” I paused then turned on the metronome. The rhythmic ticking was a bit slower than their hearts. “With each beat, your acre of forest disappears. That is what is happening today.”

A couple of weeks later, a test question asked the rate of tropical deforestation. The previous year’s class had been given the number in a lecture, and about half of them got it right on the exam. This year’s students only experienced the rate via the metronome; nearly 90 percent of them selected the correct value.

* * *

Next to saving agriculture from grasshoppers, my biggest accomplishment was saving ranchers from the USDA. The feds planned to import pathogens and parasites from around the world to suppress North American grasshoppers. They were on the cusp of releasing an Australian wasp when I used the Freedom of Information Act to acquire data from their files showing that the parasite frequently attacked a beneficial species of grasshopper in laboratory studies.

The snakeweed grasshopper eats its namesake, which is a poisonous plant that costs the livestock industry more than the grass-eating grasshoppers. I didn’t figure that I’d be celebrated for preventing a really bad idea from becoming actualized. I also didn’t figure that the USDA would call me into their star chamber and threaten my career, which they did at a national meeting of the Entomological Society of America in a small, dark conference room.

Science is objective and value-free, except when money, ego, or power is at stake; which is to say, always.

* * *

Here's how things went wrong. If you spend enough time with another living being, you begin to deeply understand it. Then you develop respect and finally a kind of love. My job was to kill grasshoppers. For awhile things worked because taking a life should be the responsibility of someone who has come into a profound relationship with the victim. Good hunters will understand. But killing becomes harder with time.

I switched to gentler studies, developing keen mathematical models of population dynamics which about ten people in the world understood and gathering wonderful bits of grasshopper ecology which about five people found interesting. Then I began writing and philosophizing about the natural world, and this felt like I was honoring the landscapes and creatures.

What I do now is even harder in some ways, but it's a good kind of hard.

* * *

"Your essay touched me deeply" and "I cried when I read about your mentor's death." Nobody ever emailed to say they were moved by one of my hundred-or-so scientific papers. Of course, I'd read the work of others to see if they had cited me. I'd track down the relevant paragraph and discover that my paper was in a string of citations used to vaguely support their own idea (meaning: they'd read my title and abstract).

But when I wrote for the public, folks would send me emails about their own stories. One fellow sent me a handcrafted wooden box as an expression of his appreciation. And then there were phone calls from rather lonely or slightly batty people who seemed utterly delighted to tell me about the intersection of my writing and their lives. For a few moments, they were happy. And that's something.

* * *

"This is insulting," she complained. My newly developed laboratory exercise on "big numbers" was not appreciated, at least by this student. However, experience showed that many of her classmates would not have mastered the difference between thousands, millions, and billions. This was quite a problem when lectures, readings, and problem sets made frequent reference to these quantities.

Now I was facing a hostile business major who regarded my efforts to engage the class in an exploration of big numbers as a demeaning waste of her time. Nothing impedes learning as effectively as arrogance. Scowling at the problem set, she grumbled, “I already know that a billion is a hundred times larger than a million."

Perhaps the largest quantity true scholars can conceive is the magnitude of what they do not know. The first step toward discovery is to encounter our own ignorance.

* * *

Here’s the hardest thing about success: quitting. Most people quit too soon, but you can stay in a fight too long. I took piano lessons until I was 12 and then I quit, too soon. I hung onto a hard-driving rationalism for too long.

Sometimes it’s a matter of quitting the right part of an endeavor. I quit being Catholic when I was in college. But I didn’t quite being religious and I’ve been Unitarian for the last 30 years (and gained respect for Catholicism).

I quit being a professor of entomology after 20 years in the field. I knew that while science was a necessary part of who I was, it was not sufficient. So now I’m working on environmental ethics, philosophy of ecology, and nature writing. That one was about right.

* * *

The coolest thing I discovered was the fate of the Rocky Mountain locust, the insect that blackened the skies of the pioneers—and then suddenly went extinct. It turns out that in their effort to feed the burgeoning mining industry in the late 1800s, a bunch of unwitting farmers armed with cows and plows destroyed the sanctuaries of the insect in the fertile river valleys of the Rockies.

But what I really discovered, for myself, was the power of story. Richard Kearney wrote: “Telling stories is as basic to human beings as eating. More so, in fact, for while food makes us live, stories are what make our lives worth living. They are what make our condition human." The locust’s story became my first, major book. And I found that while science provided me with a living, writing made my life meaningful.

* * *

For years we’d looked unsuccessfully for large-scale changes to account for the disappearance of the Rocky Mountain locust, which formed swarms equal to the area of Colorado. And then on a very long drive with my research associate, I wondered aloud, “What about an ecological bottleneck?”

“Yeah, maybe the locusts and settlers competed for the same prime real estate,” Larry took a slug of gas station coffee.

“Most of the West is hardly a gardener’s paradise,” I mused, grimacing at the bitter coffee that kept me alert.

“I’ll dig up agricultural records from the late 1800s,” he said. And then our conversation wandered into the challenges of high altitude gardening.

We found that between outbreaks, the locusts depended on the fertile valleys of the Rockies—precisely the habitats that farming utterly transformed. Sometimes, discoveries need repeated failures, meandering conversations, speculative moments, and bad coffee.

* * *

With midterm exam scores averaging 67 percent, a radical detour in teaching seemed necessary to salvage the course. The class was devoted to the science of biodiversity, and I wanted the students to derive deeper meaning from the experience. Now, however, as I sat listening to each student’s strained efforts, the idea of having them memorize and interpret poetry about the natural world seemed a failure. Until Eric—a rugged range-management major with a ragged series of quiz grades—sat down across the table, grew calm, and then serene.

With genuine understanding in his eyes and passion in his voice, Eric recited the first poem (I learned later) that he'd ever committed to memory: "…To see a World in a Grain of Sand / And a Heaven in a Wild Flower, / Hold infinity in the palm of your hand / And Eternity in an hour. …"

* * *

In 2008, my dad had prostate cancer and my mom had a stroke. During my mom's rehabilitation they contemplated what might have happened if they'd been far from home when either fell ill—and they decided to continue traveling.

The French philosopher, Jean-Jacques Rosseau asked, "Is the moment when we have to die the time to learn how we should have lived?" My dad died three years ago in a car accident, and I've been coming to terms with mortality.

I don't want to die, but even more, I don't want to find that I haven't lived. And so, I moved far from my academic home in science into the arts and humanities, knowing that in the search to discover the nature of the world and our place in it, we are far less likely to regret what we've tried and failed, than what we've failed to try.

* * *

In London, New York, and Rome millions gathered in a world record-setting-protest—and in Laramie, Wyoming, seventeen people demonstrated against the war in Iraq. The group included one of my students, who was asked by a local reporter, "What can so few people accomplish?"

The young woman replied with a sweep of her arm: "There's nobody here but us." He cocked a graying eyebrow, and she went on. "Our voices are few, but they are infinitely louder than silence." He looked at her paternalistically. "We aren't here to be nice," she said tensely. "We came because it matters."

The story never appeared in the paper. It would have been nice, but it didn't matter. We all seek the big breakthrough in politics, art and science, but my student had discovered that our lives are truly measured by the authenticity of our commitments.

* * *

My daughter wants a life in academia. She aspires to a faculty position at a top tier university, as I had. But instead I ended up happy.

She's working on her doctorate in political science at Northwestern University, and she asked for my advice on a matter of departmental politics. She knows more about politics than I can fathom, but I still have something to offer.

I never knew whether she was paying much attention to my work, but there must've been something that made her think that what I did was worth emulating. Maybe it was the freedom to pursue ideas, or the lack of supervision, or the opportunities for travel, or the passion for discovery, or my odd and lovable students and colleagues that she encountered. I don't really know, but I like that she thought there was something worthwhile about my work.

* * *

I started writing books about things that fascinated me—the disappearance of the infamous Rocky Mountain locust, the history of entomological warfare, and the ways that insects have infested our psyches. And I wrote essays about my spiritual foundations as a scientist—what it meant to kill millions of creatures, how I prayed during my days in the field, and who mattered deeply in my life. I think that I was building confidence for my current project.

I'm writing a book about how the energy industry shapes public discourse and how corporations wield their money and power to censor speech. The stories are based on events in Wyoming, but the lessons and warnings apply to all places where the accumulation of wealth relies on injustice. The book will make some very powerful people very mad.

To be honest, I'm mad too—and a little bit worried.

* * *

This weekend, the grass was cut, the garden weeded, the cat had his shots, the groceries were bought and put away, the broken trivet was repaired, and the weaving that we brought back from India was hung behind a cleverly mounted sheet of Plexiglas over a stairway that made the whole process thrilling from atop a ladder. All of this will be justifiably forgotten well before you read these words.

I'm not sure how much I matter to my wife's happiness. I'm not a detriment, most of the time. And after 31 years of marriage, I know that we love each other—and that the yard and cats and pantry and whatnot are all more important than grant proposals, committee reports, lecture preparations, and academic emails.

At least that's true on a Sunday afternoon in August.

My Accidental Scholarly Discoveries

Russell Belk

Russell Belk is past president of the International Association of Marketing and Development and is a fellow and past president of the Association for Consumer Research. He initiated the Consumer Behavior Odyssey, the Association for Consumer Research Film Festival, and the Consumer Culture Theory Conference. His research involves the meanings of possessions, collecting, gift-giving, materialism, sharing, and global consumer culture. He is currently Professor of Marketing and Kraft Foods Canada Chair in Marketing at the Schulich School of Business, York University in Toronto, Canada and holds honorary professorships in North America, Europe, Asia, and Australia. His work is often cultural, qualitative, and visual. He has published approximately 550 articles, books, chapters, and videos. Most recently he co-wrote or co-edited the books: Qualitative Consumer and Marketing Research *(2013),* Research in Consumer Behavior *(2014),* Consumer Culture Theory *(2013),* The Routledge Companion to Digital Consumption *(2013), and* The Routledge Companion to Identity and Consumption *(2013). A 10-volume edited set of his works with comments is being published by Sage.*

There are many traits that facilitate scholarly discovery, including curiosity, preparedness, persistence, observation, daring, intuition, and out-of-the-box non-linear thinking. But I have found that the most successful or well-received of my ideas have come up and tapped me on the shoulder as more or less unanticipated surprises. What I'd like to do here is to give several examples from my work and then try to diagnose what led to these pleasant surprises. The first example is an old (1989) but heavily cited paper on the "extended self." The basic idea is a simple one: we are what we own. But this wasn't the paper I started out to write. In fact I had planned to write a book focusing on materialism. In the course of writing the book (which never in fact emerged) I began on a chapter I had titled "Possessions and the Self." I sometimes feel I don't know what I think about a topic until I write about it. It was that way with this chapter.

Rather than the chapter being about how being acquisitive and possessive make us materialistic – the topic I had intended to write about – I began to think and write about how possessions can enlarge our sense of who we are. I then began to ask how it is that we come to think of things first as ours and secondly as a key part of our identity. This led me to ask what happens when we lose our possessions by accident or to theft, how incorporating other people versus possessions into our sense of self helps to define us, whether we can behave possessively toward other people, how and why we discard possessions, which possessions we feel are most a part of us, how we care for possessions that are seen as more or less central to our identity, and a variety of other questions that sprang forth.

The notion of extended self suggested itself as a key metaphor, although I recognized that it was a Western and male metaphor. So I began to look at these questions across times and across cultures. Some of my journey drew on what I thought or had read, part led me to a very diverse literature, and part sent me to the field to try to determine what possessions people actually regarded as extensions of themselves and how we might go about measuring their attachment to these things. I settled on a task to generate a list of possessions that were potentially relevant to a group of people and then had them first sort cards representing these things into "me" and "not me" piles. They then took each pile and further sorted the items into "a little (not) me" and "a lot (not) me" piles. I could then see how the item's position on this 4-point scale related to the person's reported care of these things and their willingness to give up the possession. This notion of possession attachment led me to think about different types of possessions including places, various people with whom we identify, body parts, and other non-object things that might be regarded as possessions.

Even in individualistic and independent societies of the West, the discovery that our sense of identity does not end with our skin and instead incorporates people, places, and things outside of our selves has caused me and others to think differently about identity. This revelation means that we can no longer consider identity as something that can be condensed to personality, attitudes, or other purely individual constructs. Instead, we are what we consume, who we associate with, where we live or visit. We are not only enriched by these external and experiential associations, we ARE these self-appendages. Recent work suggests that even our online avatars, photos, and posts are important contributors to our sense of self. The extended self is a very basic but powerful perspective with which to understand the shaping of our identity and its impact on our behavior.

I became so absorbed in thinking and writing about this surprisingly fruitful topic that I forgot about the book. I wrote a journal article about consumption and the extended self as well as another article about the original focus of the book: materialism. The extended self paper turned out to be my most cited work to date.

A second happy accident emerged when I was studying gift-giving – a topic that had interested me ever since I took my first teaching position. I had studied gift-giving in various contexts including dating, Christmas gifts, birthday presents, Valentine's Day gifts, and greeting cards. I had examined differences in gift-giving in Japan, China, and North America. I had detailed the characteristics of "the perfect gift," examined the gift selection process versus selecting goods for self-use, looked at gender differences in gift-giving, analyzed the meanings of Santa Claus, and investigated the relationship between gift-giving and materialism. I had learned, for example, that those who were more materialistic were less generous in gift-giving, that women bore the brunt of gift selection responsibility, and that Santa Claus and the secular Christmas help socialize children into acquisitive consumption. But I still lacked a full understanding of gift-giving. It was in considering the contrast between gift-giving and the major alternative means of product acquisition – buying in the marketplace – that I came to the realization that these were not the only ways we can acquire things. The missing piece was again obvious in retrospect. Sharing was the missing piece that had been excluded from virtually all prior work on gifts and marketplace exchange. Yet it is the oldest and most pervasive of all means of acquiring things. Before there were gifts, before money, and before barter, there was sharing. Before we are even born, our mother shares her entire self with us and this continues throughout our childhood if not beyond. Sharing remains the most common form of acquisition and distribution within the household. It is being re-invigorated outside of the household through various forms of online-facilitated sharing ranging from car sharing, to home sharing, to sharing children's toys. With this realization I began another extensive journal paper with the simple title, "Sharing." All the years I had spent studying gift-giving had kept me from seeing what was under my nose the whole time – the simple act of sharing.

Sharing, and a related for-profit business model called collaborative consumption involving short-term rental of cars, homes, rooms, and designer goods, are currently booming, thanks in large part to the facilitating effects of the Internet. Some have compared the current boom in sharing and collaborative consumption another Industrial Revolution. Even car companies like Mercedes, BMW, Volkswagen, Peugeot, and General Motors have begun to realize that selling cars to individuals is a business model that is in danger of disappearing. In order to adapt to this development, they have begun to offer cars for short term use in major urban areas as well as facilitate car owners renting out their cars and offering rides for payments. There are important implications for more sustainable consumption and business practices.

The third serendipitous discovery that I stumbled upon, again started out in researching another topic. In this case the original topic was digital consumption. I had edited a book on the topic with Rosa Llamas in Spain. But the big project was a conceptual paper. I had read over 1500 books and articles for the conceptual paper that I was working on with colleagues Rob Kozinets at my university and Henri Weijo in Finland. We had gone through two or three drafts of a lengthy paper, but hadn't reached closure on the paper. While waiting for the latest revision from my colleagues, it suddenly dawned on me that the many developments in digital consumption, including the Internet, e-mail, instant messaging, texting, blogs, forums, photo sharing sites, video sharing sites, web pages, various smartphone apps, and a growing number of social media sites, were all new since I wrote the extended self paper referred to above. Furthermore all of these new developments ushered in a host of new ways of representing our selves. I had recently co-edited another book on identity and consumption with Ayalla Ruvio from Israel. Suddenly the two topics—digital consumption and identity came together. They came from totally separate projects which suddenly merged in my mind. So I turned to writing a journal article on the topic. I was able to draw on much of the literature I had acquired for the original digital consumption paper, and my literature review also greatly expanded. The paper came together over the next several months while I was still waiting for co-author responses on the original digital consumption paper. My paper on the digital extended self was published this year and we are still seeking closure on the original paper.

The digital extended self has implications not only for how we are able to define and present ourselves to the world, but also for losing control of our identity as others post and tag us in photos, add comments about us on social media, and rate us in marketplace transactions with online services like eBay and Amazon. These known and anonymous others help to "co-construct" our identity as presented to others online. We also tend to be more disinhibited in sharing information about ourselves online than in face-to-face encounters, sometimes leading to "over-sharing" which may come back to haunt us in contexts ranging from school admission decisions and job hiring to dating and breaking up. With the greater power of the Internet to broadcast ourselves, comes an increased danger of losing some control of our online identity, prompting some to call for managing our online persona as if we were a brand.

What lessons have I learned from these experiences? Perseverance and persistence are important – it may take months or years, and in the case of the last paper twenty-five years had elapsed since my initial paper on the extended self. Preparation is all-important, even if you aren't quite sure what you are preparing for. Since I began my research and writing career I have had a broad and sustained interest in gift-giving, materialism, the meanings of possessions, and collecting. In many of these areas, like gift-giving and collecting, rather than pursuing mundane consumption objects, the importance of the object is magnified. Within these realms we give great attention to the objects we select to serve as gifts or enter our collections, not for any utilitarian reason but because of what these objects symbolize. We convey our love and impressions of the recipient when giving gifts and we form our own little world when forming a collection of objects that have special meanings to us. By selecting objects that are taken out of ordinary everyday consumption, we often find transcendent meanings in consumption objects. Such meanings of objects become meanings of life in a consumer society. They define our relationships and show what is important to us.

Over the years I have also added consumer identity, sharing, and global consumer culture to this list. This eclectic set of interests in a series of very broad topics that have been studied in a variety of disciplines led me to read very broadly. Nothing was off limits from comics to philosophy. Novels, films, observations of daily life, conversations, online news items, poems, art, and hundreds of other things all influenced me. The strength of weak ties phenomenon means that people and sources outside of the narrow confines of an avowed field of study can be especially helpful. In 2001 I co-founded a consumer research film festival in conjunction with my major conference. It is still going on and has provided a number of provocative videos that have also helped free my creativity and imagination. So what comes out of the process of discovery is not only a result of happy coincidences, but also depends upon accumulating a large array of ideas to draw upon, inspire, combine, break apart, and challenge in the process of forming and shaping ideas. Once I get an idea or a metaphor I play with it and try to track down everything of potential relevance. A key element is giving myself time for the muse to visit me – often unbidden while in a semi-wakeful state, while out running, while sitting in a conference session, while reading, when I am traveling, or at many other times when I am not "on task" and trying to come up with a thought on a particular topic. Some of these ideas are discarded, sometimes after struggling to try to bring them to fruition. But once the blockbuster idea hits me I am almost certain that it is something I can get published and that I will be truly happy with. It is different, new, and something I am willing to take a chance on.

If these are the things that have kept me searching and engaged in a research career of over 40 years, I have also been blessed with an obsessive drive to tease out the ideas that will provide real insights into a topic. Each of the three papers discussed here were over 80 double-spaced pages in draft form and included well over 200 references. But in each case I also had to be flexible enough to change topics and refocus in mid-stream. For me, obsessive persistence must be accompanied by a willingness to abandon one obsession for another that springs forth despite attempts to focus elsewhere. Flexibility and invention are then followed by a lot of hard work. Freud called the free form, childlike, playful, unintended state of creativity "primary process thinking." Only at some later point, while trying to write up the ideas or present them to others does "secondary process thinking" come into play, which is more linear, logical, and ordered. This does not mean that there is no room for creativity in writing or in other forms of presenting ideas; quite the opposite is the case. But it is only in the writing up that I bring my critical self to bear. Before and even during that stage I write a hundred or a thousand little notes to myself (before I forget them) which I will only later consider more objectively. Then I spend weeks or months revising until it feels right. I can be certain that reviewers will have other ideas and challenges, so it is not done yet, but it is my best effort at the time.

The poets have long invited their muses to visit them—a trope that suggests that discoveries come from outside of ourselves. It really feels that way when we begin to see things differently. If not a feeling of a muse descending upon us, then a curtain being lifted, the ethereal becoming concrete, or simply stumbling upon a simple but powerful insight that was there all along. As each of these descriptions implies, discovery – at least for me – comes *to* us as much as *from* us. Yes, it takes broad preparation, but no amount of disciplined reading and thinking is likely to suffice. Rather it is time apart from the project at hand and a willingness to accept the gift of the muses, even (or especially) at the cost of abandoning all we were initially seeking to do.

How I Became a Paleobiologist

Conrad C. Labandeira

Conrad Labandeira is a senior research scientist and curator of fossil arthropods at the Smithsonian's National Museum of Natural History, in the Department of Paleobiology. Conrad was elected a fellow to the Paleontological Society in 2014.

I was born in 1950 in Hanford, California, smack in the center of the agricultural Central Valley of California. As the county seat of small Kings County, Hanford at the time had a population of several thousand; it was characterized by downtown angle parking, Woolworths and Mode o'Day stores, and the renowned Imperial Dynasty Chinese restaurant. Riverdale, my boyhood town nearby, had a much smaller population of about 800 and included the primary and secondary school that I attended. The tallest building in town was the Brown Feed and Seed elevator next to the train tracks. I was raised during a time when milk from local Superior Dairy was regularly delivered to the doorsteps of rural homes in large glass bottles. The small ranch that I called home was about three miles distant from Riverdale. A slough surrounded by oaks, cottonwoods and willows often had an annual dryout. At those times I would explore the shrinking ponds that contained crayfish and small fish such as minnows, carp, sunfish, bluegill and smallmouth bass. Exploring the slough is one of the few reminisces of my early years.

My life can best be described as consisting of three phases. First were the early, formative years of working on the ranch and farm. Second, and with some overlap, were the drywall years which provided me a source of income. Third, with considerably more overlap, was my student and professional career as a paleobiologist.

Beginning during the earliest phase, my father inherited about 30 acres of land, including pasture, a meandering slough, and a dairy of about 25 Holstein cattle and a few bulls—quite small by California standards. After my dad sold the dairy cattle, and before he acquired whiteface beef cattle, a corral was built with thick, creosote-soaked railroad ties that served as anchoring end posts. These end posts were seated in amazingly difficult soil with hardness seemingly of concrete. A hole about 3.5 feet into the ground had to be dug with a post-hole digger, which took about two months using ample water poured into the hole to soften the hard-pan layer, and repeating the procedure week after week. Once the corral was built and beef cattle were acquired—the dairy cattle were sold—the enclosure was used for branding, castrating bulls to turn them into steers, and treating cattle maladies, such as use of a sulfa-drug powder to eliminate pink eye, a common ailment. All these activities involved use of the chute and stanchions on one side of the corral. I recall one experience that may have nearly cost me my life—I must have been about 11 or 12 at the time—in which an unhappy horned bull came loose from the stanchions during a pink-eye treatment and came after me in the center of the corral. I ran to the wood rails to hop over the fence just in time to avoid being gored.

Soon beef cattle ranching became a losing enterprise, and my father started planting crops on the same acreage. The crops initially included permanent pasture and alfalfa for hay. The alfalfa involved considerable tractor work that pulled scythe for cutting, a rake forming hay rows, and baler for packing rectangular bales of hay.

Once the bales were in the field, we contracted a hay-hauling service to collect the bales and transport them to other ranches for sale. Other crops included corn and barley. One crop that loomed especially important for me was cotton, principally because we lacked a converted John Deere tractor for mechanical cotton picking, and the cotton had to be picked by hand. (I suspect that it was cheaper to pick the cotton by hand, as we were paid piecework by the pound, but I never questioned my father about this.) With both hands moving quickly and synchronously, white fluffy cotton was picked from the bolls; a process that often I ruined by grabbing dirty cotton from the ground. The overall process involved pulling a long cotton picking sack along a furrow on the ground that was made of thick canvas and had a padded shoulder strap that ran diagonally across my chest. I was surrounded on both sides by two, tall rows of cotton plants; occasionally the cotton had to be shaken and packed toward the end of the sack. At the end of the row I would make a U-turn and continued on the back side of the recently partly picked row. When the sack was full, it was taken to a tripod with a scale and the sack was weighed. I was paid on the spot—it seemed to me a pittance—but the job was only completed after I climbed the ladder along the side of the chickenwire trailer into which the cotton was dumped. At the end of the day, I rode with my dad on an old, orange Case tractor to the cotton gin, about 10 miles away.

I clearly remember, almost as if it happened yesterday, a few experiences as I was picking cotton that captivated me and was more important than either taking a short snooze on the cotton sack under the 100 degree plus heat or picking the cotton for financial reward. I noticed that the undersides of the leaves and the leaf-like appendages at the base of the bolls contained a marvelous world of insects. Particularly fascinating were tiny, delicate, lime-green, soft-bodied insects that inserted a needle-like structure into these leaves and stems. These insects would remain stationary for considerable time, but were harassed by the activities of, as I learned later, the black-and-orange larvae of ladybird beetles. The beetle larvae would often hold their aphid prey aloft with their mouthparts as the insect prey squirmed. I wondered to myself—I was 11 or 12 at the time—how fascinating these interactions were and whether there were people who studied these insects, even though the entire idea of research was totally foreign to me.

I continued assisting my father on the farm during my completion of grammar and high school. During weekdays, study time was after dinner.

My father and mother kept the ranch going, but dad also got a job in the drywall construction business. He became a journeyman taper, but also did finishing and occasionally worked on a spray rig that applied wall texture and ceiling acoustic. Initially dad was involved in work along coastal Southern California in places such as Santa Barbara, Ventura and Santa Paula that were undergoing a construction boom, returning home late in the evening on Friday and departing for work on late Sunday afternoon. This routine continued until he landed a job in Fresno, twenty-four miles away.

I would work with my father on new residential homes, remodels of older homes (such as garage conversions to family rooms) and commercial construction on weekends, holidays, and the Christmas, Easter and summer vacations. Wages were infinitely better than working the farm, although my father initially provided me one dollar per week as an apprentice to him. Within a few years I became a journeyman taper–finisher during the summer before I graduated from Riverdale High School. During that summer there was a need for me to be the breadwinner, as my dad recently had major back surgery from driving an early 1920's large Case tractor that required backbreaking effort in steering, braking and especially in turning a hand crank that had a nasty kick.

Later, I worked for various contractors as I attended Fresno State College—later to become California State University, Fresno (CSUF)—as an undergraduate. It took me 11 years to pursue three majors (biology, geology and anthropology) and two minors (Latin, Physical Science), which provided me a refuge from the world of agriculture and drywall. Nevertheless, I used drywall as my sole source of income that allowed me to pursue college. I often arrived in class with fine white power covering myself and clothes, resulting from the lack of time to change my clothes in some closet or outhouse at the nearby construction sites.

The world of drywall was quite different from my days on the farm and the world of undergraduate education at CSUF. The most challenging times were the changeovers from the end of spring college classes and the beginning of drywall work during summer "vacation," and the return back to college at the start of the fall semester. It was at these times that my language changed, and embarrassing times occurred when more colorful drywall argot surfaced in class discussions at CSUF, and conversely, the use of fifty cent words were misunderstood by my drywall compatriots. Much of my work was done in Fresno or nearby towns such as Clovis, Sanger, Selma, Reedley and Kingsburg, but also in the Sierra Nevada foothills. For those jobs in the mountains, I was afforded a rare opportunity to wander into and observe diverse plant communities of manzanita, California lilac and buckbrush forming chaparral shrub, the blue oaks, digger pine and California buckeyes of adjacent woodland, and the special, unique vegetation of serpentine-rich soils. I would collect leaves with insect damage and identify the plant hosts and insect herbivores that made the damage. Just spending an hour in the surrounding plant communities after the job was completed and the drywall tools and buckets were loaded into my pickup was worth just as much as the piecework money that I made from the homes.

Without sounding too self-serving, I was—for about several years—the fastest taper, also known as a bazookaman, in the Fresno area. I was a bit slower at finishing, and never got the knack of dragging a hose line for spraying texture and acoustic. However, this newfound, quick, skill eventually allowed the other drywallers that I worked with to stop making fun of my listening to National Public Radio, or reading *Scientific American* during the lunch breaks, or questioning my sanity as to why I would want to return to the University at the end of summer. I was often reminded that college was for people with pointy heads and one could make so much more money doing drywall as a livelihood.

One day during the late 1970s I entered my paleoecology class at CSUF, and the instructor, whom I knew quite well, said that he wanted to see me after class. He looked quite perturbed. As I sat down in a chair facing him, he looked straight into my face and said to me, "You need to know what you want to do in life," and suggested that "I had a choice," indicating that I should "move on." At first, I found this shocking, but upon reflection I realized that I had been at CSUF for way too long and my stay there had become too comfortable—particularly as I was taking seemingly every course for which I had interest in. Soon, I took Graduate Record Exams in geology and biology and applied to 32 graduate schools in the U.S., Canada and England.

After receiving some offers and many rejections, I settled on the University of Wisconsin at Milwaukee (UWM). At UWM my advisor was Dr. Peter Sheehan, a fellow Central Valley Californian as it turned out, from the city of Lodi. We settled on a Master's thesis investigating the systematics, taxonomy and ecology of the 490 million-year-old (Late Cambrian), shallow marine trilobite *Dikelocephalus*. I worked out the systematics of this genus of shallow-water-inhabiting trilobites and in the process eliminated about 15 or so taxonomically unjustified species that were published in a report during the 1930's. The process required me to collect specimens in the field, along the Wisconsin River of south-central Wisconsin, and up the St. Croix River, to the north that forms the boundary between Minnesota and Wisconsin. Most of the trilobite material had been collected decades earlier and were housed in the collections where Peter worked, in the Milwaukee Public Museum (MPM). At the MPM I had a desk where I did all of my morphological measurements and analyses for my thesis. I conducted numerous measurements on the head shield and lower mouthplate, the thoracic segments and the tail shield of many, mostly fragmentary *Dikelocephalus* specimens. The measurements were later plotted in univariate, bivariate and multivariate ways and then published as a thesis and later as a report in the *Journal of Paleontology*, together with fellow graduate student Nigel Hughes, from Bristol University in the United Kingdom, who also was working on related trilobites in the same region.[1] In my concluding thesis chapter, I also had something to say about changing species concepts in American paleontological thought since the 1920's.

The master's thesis that I did at UWM would play a major role in the next step in my academic life. My experience in working out the species limits of *Dikelocephalus* specimens was my first taste at morphological data, and I found out that the resulting plots of measurements were highly robust and exact—they followed fairly straight lines with little scatter of points–and impressed upon me the well-behaved nature of measurement data taken from organisms. These data and analyses were in distinct contrast to the considerably messier results of ecological data that I was to experience in my postdoctoral research and my current job. I gradually learned to not only tolerate the ill-behaved nature of ecological data over morphological data, but even began to find ecological data more interesting and intriguing. This and other experiences allowed my UWM graduate student years to expand my research competence in many ways, different from that of CSUF. This validated my CSUF paleoecology professor's admonition that I had needed to move on. My time at UWM was the ideal prelude to what came next: pursuing a doctorate at The University of Chicago.

Peter was a colleague of Dr. John ("Jack") Sepkoski at The University of Chicago (UC). Toward the end of my matriculation at UWM, in March of 1982, Peter took me to a symposium sponsored by the Fishbein Center for the Study of the History of Science and Medicine at UC, where I became reacquainted with Jack. The three day-long symposium, *Persistent Controversies in Evolutionary Theory*, involved topics with follow-up discussions such as the "Reality of Group Selection," "Mechanisms of Speciation and Modes of Change," "Religion, Morality and Evolution" and "Social and Behavioral Evolution," with speakers that soon were to form a major influence on the evolution of my own thoughts at UC, including Dr. Tom Schopf at UC and Dr. David Raup, then at Chicago's Field Museum of Natural History. Many of the ideas presented just blew me away and took me back to the readings and class discussions of my early CSUF years. The ideas presented in the symposium involved looking at the fossil record in different ways and how to interpret evolutionary biological data. At the urging of Peter, I had already applied to UC's doctoral program in Paleobiology, together with the six other universities. Although I wasn't set on one particular doctoral program, I was urged by Jack Sepkoski to visit UC and interview as a prospective student.

On the day of the interview, I first met with Jack, and we had a lively chat about our common interests in Cambrian trilobites, their depositional environments, and my earlier interest in insects. Jack suggested additional ways that I could test the conclusions of my thesis on *Dikelocephalus*, although he agreed with the conclusions. He was keenly aware of the problem with "layer cake" stratigraphy that was the reigning paradigm during the 1920's to 1940's that led to the useless multiplication of *Dikelocephalus* species that I subsequently discarded at UWM. After a pleasant exchange of views and presenting arguments as to why I should attend UC, he looked at his watch. He immediately grabbed me by the hand, and almost threw me into Tom Schopf's office next door, saying to Tom, "Here's Conrad," and then he closed the door.

I never have had a meeting as initially traumatic as I had with Tom. As I composed myself after entering the office, Tom barely acknowledged me, saying something like "Oh, you are the new doctoral student prospect," as he was looking down at his mail and tearing the margins of envelopes with a letter opener. After a few additional comments, including a few inadept ones from me, the tenseness of our encounter became tangible. Still looking at his mail, Tom said, "So … tell me about your thesis." I proceeded to provide the intellectual backdrop to the project, stated the methodology that I used, and related the results regarding my taxonomic revision of *Dikelocephalus*, and then I concluded in an impromptu presentation of how there has been a dramatic change on the use of species concepts by paleontologists since 1930s, when the *Dikelocephalus* species I was studying were initially described. Tom abruptly cut me off and said "What makes you think you used the correct approach in establishing your revision of these trilobites?" I wimpishly attempted to state the reasons why the initial describers of *Dikelocephalus* in 1933 misinterpreted the stratigraphy that contained the trilobite specimens and related this error to how paleontologists conducted taxonomy. This was followed by fusillades of counterpoints to my points, which seesawed between Tom and me for what seemed like an interminable time. Somewhat in exasperation, I asked Tom what specific techniques *he* would have used to resolve *Dikelocephalus* taxonomy. There was a short period of silence, and he mentioned a technique, to which I countered that it had been used for the taxonomy of another group of fossils and was found wanting. After an interim of progressively less unfriendly banter, Tom led me to his office door, and said to me in a hushed tone, "I think you will do well here," and closed the door. Then Jack took me to interview with Prof. Alfred Ziegler, a paleogeographer, some of whose work I already had known. I was in more familiar territory of country music in the background, a round case of Copenhagen snuff in Fred's shirt pocket, and the occasional burps and other incivilities. I enjoyed Fred's frank observations and we had a great discussion of current projects in Fred's lab.

The years at UC were the best intellectual environment that I have had. Once I matriculated at UC, Prof. Schopf took me—I was the only student he did this for—to various orally read papers among UC faculty such as Robert Richards, and visiting colleagues such as Garland Allen and William Wimsatt. The readings were irregular meetings of several professors on the Committee on Conceptual and Historical Studies of Science, of which Tom was a member. After one of these meetings, one exchange that I never have forgotten is a question that I asked Tom as to what is important in becoming a scientist. He unhesitatingly responded that: "It takes three qualities. First, you need to be smart. Second, you need to have a vision of where you want take the discipline. And third, you need to have considerable ambition and drive. And, of the three, intelligence is least important." I never forgot this discussion, and it stood in stark contrast to lighthearted moments such as pranks done to students and to faculty alike. In one instance my officemate rearranged all the books of my extensive library based on the sequential colors of the light spectrum, resulting in my inability to fine any volume formerly listed alphabetically by author. Another prank turned the tables on a sedimentologist in which hundreds of large balloons were inflated and placed in his office, where they occupied the upper third of the room.

Toward the end of my stay at UC, I was working feverishly on my dissertation, which was a massive undertaking of morphologically coding 1,365 species of modern insect mouthparts that represented 70% of all insect taxonomic families which were analyzed with 49 mouthpart-related variables. The result was a phenogram that delineated the 34 fundamental mouthpart types of modern insects. Each mouthpart type was tracked through deep time of the geological record to determine which lineages of insects possessed a particular mouthpart type, how ancient was the particular mouthpart type, and whether there were multiple origins of the same mouthpart type among different major insect lineages. When all 34 modern mouthpart types were plotted through time, and allowing for reliable to less reliable rendering of the data, it turned out that most mouthpart types are quite ancient, extending to mid-Mesozoic, approximately 200 to 250 million years ago, and seven mouthpart types—including two extinct mouthpart types—extend very early to the late Paleozoic, originating between 410 to 320 million years ago. This pattern is perhaps better understood when the diversity of the 34 modern and two extinct major mouthpart types were plotted through time. It appears that there are five major phases of mouthpart innovation, and more than half originated by the beginning of the Mesozoic 252 million years ago. It also is notable that mouthpart diversity outstrips insect biodiversity (at the taxonomic family level) at least during the Mesozoic and Cenozoic, possibly earlier.[2] The dissertation also involved many gory details of the intricacies of mouthpart elements morphology—labrums, hypopharynxes, mandibles, maxillae and labia.

There also was a major provision in the dissertation for examination of Paleozoic insect mouthparts from the famous Late Carboniferous Mazon Creek deposits in northeastern Illinois. As the time of my defense approached, I contacted Prof. David Raup, my advisor, and we discussed the thesis. David remarked that I need to finish up soon, and then brought up the Mazon Creek insect mouthparts substudy. I told him that I was quite enthused about relating physical fossil evidence with the phenetic analysis. He then straightforwardly informed that "the Mazon Creek mouthparts study would make for a great postdoctoral study." I was flummoxed but also quite ticked off as I walked back to my office. I really wanted to study the actual fossils of Mazon Creek insects, as an antidote to spending hours poring over insect mouthpart treatises and coding characters for mandible shapes, unique structures occurring in the maxilla that formed elongate siphons for sucking nectar, hypopharynxes designed for impaling prey and labra that formed extensive fans for sieving plankton. Examining Mazon Creek insects allowed me to get my hands dirty in the lab, but also provide a direct link to the fossil record, instead of spending time in the Crerar Library collecting character-states. Nevertheless, after mulling over the issue not including the Mazon Creek mouthpart study for two weeks or so, I concluded that David's advice on the matter was spot on, although I never informed him of my subsequent views on the matter.

As the days approached to produce my dissertation, there were major problems in running the mega phenetic analysis, after submitted job after job on the University's gargantuan mainframe computer bombed. I visited the offices of the computer advisors countless times: the format statement was correct; the variables were all appropriately defined; the BMDP syntax was fine, but the jobs kept on bombing. As my defense loomed, and my nervousness increased, I had a session with several of the IT advisors. They agreed to reset the entire UC mainframe computer buffers and for me to enter the Computation Center at 2:00 am on a Sunday and submit the job. I did this, arrived at the Computation Center output room across the campus the next morning, and was surprised to find a nine inch thick, large fanfold output with light green and white lines running across and printer sprocket holes on the perforated sides. I couldn't believe it: the immense analysis worked! I found out later at the Computation Center that my data matrix was too large for the computers, given the cluster analysis programs that I was using. The temporary (diagonal) matrices on 49 x 1365 original input matrix had to be continually and simultaneously swapped by computation algorithms and could not be handled by the mainframe computer settings.

Within one week the major task was to transform the phenogram output results from the fanfold stack to a single, readable phenogram chart. To do this, I photocopied reductions of the fanfold sheets and then stitched the resulting trimmed sheets into scotch-taped panel strips, and then scotch-taped the panels into a final assembly. It took the space of a church basement for me to assemble the panels. I then I plotted the extensive triangular phenogram measuring four by five meters on a single page. However, this was not done in time for my thesis and I resorted to presenting the thick fanfold output to my Dissertation Committee. Miraculously, my Dissertation Committee approved. I finished the dissertation, consisting of five volumes and 1186 pages, and I was awarded the degree in 1990.

During my last year at UC, I applied for a postdoctoral position at the University of Illinois at Urbana–Champaign (UIUC), 110 miles south of UC. I interviewed with Dr. Tom Phillips in the Department of Plant Biology, who would be my mentor, but also visited faculty in Entomology and the South Farm where there were row crops and experiments involving insect pests. I liked the atmosphere of the new facilities, students and faculty, because it was quite a change from UC. UIUC had significant programs in Entomology, Plant Biology and Agriculture—the latter something that was not present at UC and reminded me of my days back in the Central Valley. What really blew me away was the immense permineralized coal ball collection that had been assembled by Tom Phillips, his predecessor, and student paleobotanists in the South Farm, not too far from the swine unit and endless acreage of corn and soybeans. The UIUC coal ball collection, the six or seven rock saws, and the acetate peel archive was to me an amazing facility representing twenty million years of Late Carboniferous coal-swamp forests.

These coal-ball peels contained an exquisitely preserved record of insect damage on plants and their identifiable tissues in the form fossilized fecal pellets. This opportunity resulted in a sea change from my former research of understanding the morphology of insects that created the feeding damage, to now actually recording the types of feeding damage inflicted by insects on extinct plants from 320 to 300 million years ago. A series of papers resulted from the UIUC studies,[3] which put me in good stead for the next major shift in my career—a possible job at the Smithsonian's National Museum of Natural History (NMNH) in Washington, D.C.

After I applied for the position at the NMNH, I gave a job talk and almost was rejected outright. One USGS paleobotanist was unsatisfied with my presentation, and so I was invited for a second job talk. The considerably revamped, second presentation was much better, and my job started in December of 1991, followed by a probationary period of three years. Three years was the standard length of time before a candidate was either let go or converted to permanent status. Nevertheless, my dossier, on which the tenure decision hinged, was submitted to the Associate Director for Science and Collections at the end of two and two-thirds years, which gave me little time to get my scholarly act together. Fortunately, I had been working while at UIUC with Jack Sepkoski on writing up a succinct overview of my mouthparts work as well as an entirely new dataset on the distribution of insect taxonomic families in the fossil record. These two insect diversity aspects of the fossil record, one morphologic and the other taxonomic, was published in *Science* during 1993[4] with me as first author, which did not hurt in securing the NMNH job.

I gradually increased my productivity in research, collections and outreach following this slow fuse of a start. I soon became entangled in committee work, and at one point served on eleven departmental, museum and institutional committees. I eventually sent a memo to the departmental chair withdrawing from most of the committees, with a goal of focusing on my research and the fossil arthropod collections in my care. During my time at the NMNH, I have been fortunate to have been associated with many talented, collaborative colleagues, most outside the museum, and especially have been blessed with a string of very gifted and productive students and postdocs, such as Peter Wilf in 1999, with whom a new subdiscipline was conceived and implemented based on ferreting out fundamental types of insect-mediated plant damage present in the fossil record, currently being upgraded in the fourth version of the *Guide to Insect (and Other) Damage Types on Compressed Plant Fossils.*[5]

The *Damage Guide* has become the standard handbook for characterizing and analyzing fossil plant damage. Because of an extensive collaborative interest in the fossil record of plant–insect interactions from the Devonian to the recent, I have been involved in related research projects on all continents except Antarctica. Work on the Museum's exhibits also has kept me busy, particularly Butterflies and Plants: Partners in Coevolution exhibition (2008 - http://butterflies.si.edu), and most recently, an upgrade of three paleobiology halls in the Deep Time exhibition (http://tinyurl.com/sideeptime), the largest and costliest exhibit commissioned by the NMNH (2019). Recently, I have been involved with colleagues from Europe in examinations of new groups of plant-associated insects and their gymnospermous plant hosts during the mid Mesozoic 165 to 110 million years ago. These investigations included discovering a new, unique pollination style in the fossil record involving thrips pollinators and their cycad or ginkgo hosts during the expansion of angiosperms. Another exciting discovery was the presence of long-proboscid scorpionflies that evidently were feeding on nectar-like fluids on a variety of gymnosperms[6], that consisted of three, probably separate lineages. The scorpionfly lineages, together with five other lineages of other insects, have convergently evolved a long proboscis from varied mouthpart elements to allow pollination of gymnosperm ovulate organs with deep-throated or other tubular or channeled structures to allow uptake of nectar-like fluids.

Currently, the Paleobiology Department's Labandeira Lab has seven undergraduate and graduate students, postdocs and research associates. I feel very fortunate to be a research scientist and fossil arthropod curator at the NMNH and would not trade this job for any other.

The exploration, discovery and excitement of how I became a deep-time paleobiologist involved several "precepts" that I developed from inner resources or came to understand as continued through rather circuitous journey in attempting to understand the world around me, particularly the world of natural history. Linked to each of these perspectives are brief personal reminiscences, provided below in approximate chronological order from the time I was a young kid on my parent's ranch to a professional in mid-career.

•*Pursue curiosity wherever it may lead you.* My interest in natural science probably began in my cursory investigations of the fish, crustaceans and other dead organisms that I found in the dried-up ponds in the slough on my parent's ranch.

•*Hard work and vision are as important as brains.* My drywall experiences in funding my undergraduate education taught me that intellectual achievement in a field—at least for me—could only have been brought about by having the vision of what you want to do as a future scientist and doing the hard work to achieve it.

•*Treat setbacks as opportunities.* At the time, a major disappointment was my inability to do the anticipated study of Mazon Creek insect mouthparts as part of my dissertation at UC. In hindsight, I was able to finish the dissertation at a reasonable clip, and move quickly onto a postdoctoral fellowship at UIUC. The UIUC postdoctoral fellowship opened new research horizons that undoubtedly would not have been available had I taken the extra year to complete the Mazon Creek mouthpart project at UC.

•*Center your research on big-picture questions.* One option for graduate students is to work on a dissertation that involves descriptive research on a relatively small, well-constrained problem. Alternatively, I tackled, probably naively, a much larger issue involving a typology of morphologically complex suites of insect mouthparts, assessing the major patterns of mouthpart evolution through time, and involving the most diverse, terrestrial metazoan group of all time. The latter, more hypothesis-driven approach opens a very broad window of research activity that can last for a career and beyond and also provides discipline-wide flexibility to pursue new questions in related fields.

•*Revel in data.* Data are immortal whereas the interpretations of those data typically are ephemeral. For example, my colleagues and I have collected extensive ecological data that was used for constructing the most well resolved food web of all time, the Eocene Messel Lake food web of central Germany.[7] The food-web raw data that were posted on-line in two websites[7,8] are now being used by other labs to address issues we were unable to address at the time our food-web manuscript was submitted. This contributes to the scholarly betterment of the discipline.

•*Imagine innovative approaches.* Production of the *Damage-Type Guide*[5] was initially an effort to standardize and categorize all of the distinct types of insect damage for use in my lab and colleagues' labs. However, the uses of the *Damage Type Guide* has launched, unpredictably, a new subfield within paleobiology that assesses insect herbivory both on bulk fossil floras and on individual herbivorized plant hosts both in time and in space. Approximately 100 papers, as of this writing, have been written by various colleagues that use the new method of determining insect herbivory in time and space.

Links to Labandeira's research

Labandeira home page - http://tinyurl.com/silabandeira

Video: National Geographic Society produced documentary, Clues to the End-Permian Extinction, *with Labandeira and South African colleague Rose Prevec - http://tinyurl.com/endpermianextinction*

-

Video: National Museum of Natural History: Sheer, Abject Curiosity. *Discusses the marvels that can be discovered underneath a forest and how curiosity drives science - http://tinyurl.com/sheerabjectcuriosity*

-

Video: National Museum of Natural History: The Wrong Side of the Planet. *Discusses the role of insects in the pollination of Mid-Mesozoic gymnosperms - http://tinyurl.com/wrongsideplanet*

-

Science Story: Cockroach Fossils Found in Colorado Shed Light on Insects' Evolutionary Timeline*; interview of Labandeira by Laura Poppick of LiveScience - tinyurl.com/cockroachfossils*

-

Video: O Fungo de the Last of Us—Nerdologia 48. *Overall review of the article in Current Biology on Eocene zombie ants from the Messel Formation of central Germany (in Brazilian Portuguese) - http:// tinyurl.com/nerdologia48*

Footnotes

[1]The results of my master's thesis eventually were published in:
Conrad Labandeira and Nigel Hughes: "Biometry of the Upper Cambrian trilobite genus *Dikelocephalus* and its implications for trilobite systematics." *Journal of Paleontology*, vol. **67**, pages 492–517 (1994).

[2]The major results of my doctoral dissertation were published in:
Conrad Labandeira: "Insect mouthparts: ascertaining the paleobiology of insect feeding strategies." *Annual Review of Ecology and Systematics*, volume 28, pages 153–193 (1997).

[3]Such as the following publication:
Conrad Labandeira and Tom Phillips: "A Carboniferous petiole gall: insight into the early ecologic history of the Holometabola." *Proceedings of the National Academy of Sciences*, USA, volume 93, pages 8470–8474 (1996).

[4]The relevant publication was:

Conrad Labandeira, and J. John Sepkoski, Jr.: "Insect diversity in the fossil record." *Science*, volume 261, pages 310–315 (1993).

[5]The widely used *Damage Guide* is found at: http://paleobiology.si.edu/pdfs/insectDamageGuide3.01.pdf; the citation is:

Conrad Labandeira, Peter Wilf, Kirk R. Johnson and Finnegan Marsh: *Guide to Insect (and Other) Damage Types on Compressed Plant Fossils (Version 3.0–Spring 2007).* Smithsonian Institution, Washington, DC, 26 pages (2007).

[6]The referred citation is:

Dong Ren, Conrad Labandeira, Jorge A. Santiago-Blay, Alexandr P. Rasnitsyn, ChungKun Shih, Alexi Bashkuev, M. Amelia Logan, Carol L. Hotton, and David L. Dilcher: "A probable pollination mode before angiosperms: Eurasian, long-proboscid scorpionflies. *Science*, volume 326, pages 840–847 (2009).

[7]The citation for the food-web manuscript, and relevant supplemental data is:

Jennifer A. Dunne, Conrad. C. Labandeira, and Richard Williams: "Highly resolved middle Eocene food webs show early development of modern trophic structure after the end-Cretaceous extinction." *Proceedings of the Royal Society B (Biological Sciences)*, vol. 281, article number 20133280 (2014).

[8]The extended dataset for publication 7 above is found at the following online digital data repository:

Conrad C. Labandeira and Jennifer A. Dunne: Data from "Highly resolved middle Eocene food webs show early development of modern trophic structure after the end-Cretaceous extinction." *DRYAD* Digital Repository (doi: 10.5061/dryad.ps0f0)

What Rotifers in the Gulf of Mexico Ecosystem Taught Us About Toxicity of Saltwater Oil Spill Cleanup Techniques

Roberto Rico-Martínez

Roberto is currently a Professor at the Chemistry Department of the Autonomous University of Aguascalientes in Mexico. He was born in Celaya, Guanajuato, Mexico in 1964. He received a Bachelor of Science degree from the Autonomous University of Aguascalientes in 1987; a Master of Science degree in Limnology and Oceanography from the University of Wisconsin-Madison in 1989; and a Ph. D. in Applied Biology from the Georgia Institute of Technology in 1995. He has published more than 60 scientific articles mainly in the topic of aquatic toxicology.

It all started in May 2010. I received a pamphlet of the Fulbright/García-Robles Scholarship and decided to apply for it. I started exchanging e-mails with my brother Ramiro (a former Fulbright/García-Robles Scholar), who convinced me to apply for the scholarship. This scholarship is only given to three Mexican citizens every year with outstanding research projects. The most important part of the process (if you want to be successful), is the research project. I had been working on aquatic toxicology related topics at my university in Aguascalientes since 1996 after receiving my Ph. D from Georgia Tech.

The years that I spent at Georgia Tech earning the degree taught me the beauty of the behavior of the animals to confront the many challenges that the environment provides for them. I was interested in the myriad of strategies organisms employ to survive in the settings of new conditions and compounds released in the environment. Having completed a thorough ecological education from Wisconsin the focus on the effects of anthropogenic (man-made) substances on organisms came naturally to me. By that time, I was already in love with rotifers, these small invertebrates that I have admired since my early years as a Biology student at the Autonomous University of Aguascalientes, in Mexico. Ecotoxicology is a discipline that combines the observation of animals with some chemical principles and allows the assessment of the organism's response to these chemical or environmental samples. It also allows you to visit the field frequently in search of new specimens to culture and to assess the effects of some contaminants. This combination of field and laboratory experiments resulted in a satisfactory experience in my early years as scientist.

I approached my former Ph. D. academic advisor Dr. Terry W. Snell at Georgia Tech, and he updated me on the Gulf of Mexico oil spill that was a hot discussion topic in the U.S. at the time, and now. Basically, an "April 2010 oil spill in the Gulf of Mexico discharged 4.9 million barrels of crude oil from the Macondo well and more than 1 million gallons of the oil dispersants Corexit 9527A® and Corexit 9500A® were applied to the sea surface, and more than 770 thousand gallons to the sub-sea." This large scale application of oil dispersants, motivated us to examine the effects of the dispersants on toxicity.

Terry was excited because the U.S. Environmental Protection Agency (EPA) had selected the *Brachionus plicatilis* acute toxicity test as the one to be used by British Petroleum (now BP p.l.c.) to monitor the toxicity at the Gulf of Mexico. The toxicity test with the euryhaline (an organism that lives in a wide range of salinity) rotifer (invertebrate organism) *Brachionus plicatilis* was one of the successful projects developed by Terry as a researcher. The test consists of exposing newly born organisms (less than 24-hours-old) hatched from cysts (sexual eggs or embryos) of the rotifer *Brachionus plicatilis* to a series of high purity toxicant concentrations or to environmental samples to analyze the mortality response (number of dead animals) of the rotifer after 24 to 48 hours of exposure in controlled conditions (usually a bioclimatic chamber at 25°C).

The test provided the opportunity of analyzing thousands or even millions of environmental samples in a few days. He had access at the time to some of the results and the directives of the EPA regarding the toxicity analysis in the Gulf. We were aware of some of the limitations that many toxicity analyses in marine systems have, especially when analyzing the interaction between the petroleum and the oil dispersant, so a project started to take shape in the next few months. We sought to compare the toxicant sensitivity of *Brachionus* species (Rotifera) to develop better models for assessing toxicity in marine waters. By September 2010 the project and the application for the scholarship were ready.

I was expected to start my sabbatical leave from the Autonomous University of Aguascalientes in January 2011. I was still waiting to receive an answer from the scholarship committee after a personal interview in Mexico City in December 2010. I remember telling my wife (who did not want to spend one year in the U.S., that it was quite unlikely that I would obtain the scholarship since some of the "holy cows" of the Mexican research system (from the most important research centers in Mexico like UNAM, IPN, CINVESTAV, and so on) have applied for the scholarship and I came from a small State University in Central Mexico. To my own surprise and that of my wife, I obtained the scholarship! I started the painful process of convincing my wife to come to Atlanta for at least ten months, and to request of the authorities of my university to delay my sabbatical stay for six months. I prepared to spend eleven months from July 2011 to May 2012 in Atlanta, Georgia.

I bought a brand new small car in February 2011 in Aguascalientes. My three children, my wife, and I took a three-day journey driving from Aguascalientes to Atlanta. The first day we reached Chihuahua in northern Mexico, and the second day we arrived in Pecos, Texas after a small stop in El Paso, Texas to cross the border and change some pesos for dollars. The third day we crossed the mighty Mississippi River, cruised past Birmingham, Alabama, and made it to Atlanta.

I started getting involved with the research project immediately. We asked for oil dispersants from several companies and got a poor answer from many of them. However, we were fortunate enough to obtain a small amount of Macondo well oil (the Macondo well is the place where the oil spill originated in the Gulf of Mexico), and of 9500A® (the oil dispersant most used in the Gulf of Mexico in 2010). Oil dispersants work like the detergents that we used to get grease off when we are washing dishes, perhaps with much less toxicity than the usual formula contained in dish detergents. Oil dispersants are necessary to prevent oil for reaching the sea shoreline where its adverse effects would be worse.

The problem is stated simply in the introduction of the scientific article that we published later in November 2012: "In April 2010 oil spill in the Gulf of Mexico discharged 4.9 million barrels of crude oil from the Macondo well. One of the first responses was to apply more than 1 million gallons of the oil dispersants Corexit 9527A® and Corexit 9500A® to the sea surface, and more than 770 thousand gallons to the sub-sea. This large scale application of oil dispersants, motivated us to examine the effects of the dispersants on toxicity, especially given the limited toxicity information that is available." Our project was planned to analyze in the laboratory the toxic effects of Macondo oil and Corexit 9500A® and the potential synergistic effects (if any) of this combination. We knew that oil dispersants have been used widely, but there are doubts in the regulatory community about the efficacy of dispersants to reduce the biological impacts of oil spills because of the poor understanding of oil dispersant toxicity. This problem is particularly complicated in the marine system where different complex mixtures are formed between sea water and chemical contaminants. The petroleum oil is in fact a complex mixture of many substances with different properties and the solubility of each component varies widely. Finally, it is well known in some cases and suspected in others, that oil dispersant tends to increase the solubility of the oil mixture and therefore increase the bioavailability and toxicity of the toxicants that conform the oil mixture.

The months passed by rather quickly in the laboratory of Dr. Terry Snell at Georgia Tech.

We obtained different types of euryhaline rotifers from several parts of the world including some from Veracruz Mexico, sent by Dr. Alejandro Pérez-Legaspi a friend and former doctoral student of mine. The goal of the research was simple. We tried to answer the question: What happens when you mix Macondo oil and oil dispersant (in this case Corexit 9500A®) in terms of toxicity (adverse effects)?

The answer we got was amazing!!!

"When Corexit 9500A® and oil are mixed, toxicity to *B. manjavacas* increases up to 52-fold. Extrapolating these results to the oil released by the Macondo well, suggests underestimation of increased toxicity from Corexit application." Simply put, the oil dispersant made it 52-times worse the adverse effects of the oil for this particular species that is known to inhabit the Gulf of Mexico and that was picked by EPA as a representative of the organisms. Many scientists would argue that this is only a result from laboratory experiments and that a complete field study might be necessary to better assess what really went on in the Gulf of Mexico. True. Other scientists might argue that we did not perform a comprehensive chemical analysis of the oil components during the toxicity testing. Also true. However, that does not weaken our argument related to the increase in bioavailability caused by the oil dispersant that resulted in a huge increment in the adverse effects caused by the oil to B. manjavacas, and many other species inhabiting the Gulf of Mexico at the time of the oil spill.

You might notice the change in the species name of the rotifer from Brachionus plicatilis to Brachionus manjavacas. That is just other finding of our research. Although registered originally as Brachionus plicatilis toxicity test, in reality, we know the species used in the test corresponds to Brachionus manjavacas according to recent genetic analysis (which was included in our scientific article).

The important thing is what the heading in the Georgia Tech website by Jason Maderer says: "Gulf of Mexico Clean-Up Makes 2010 Spill 52-Times More Toxic - Study shows mixing oil with dispersant increased toxicity to Gulf's ecosystems" (http://tinyurl.com/gulftoxiccleanup). The attention we obtained from this post is related to the fact that when extrapolating these laboratory experiments results to the whole Gulf of Mexico ecosystem, the assumptions are scary. The lack of information about the adverse effects of the synergy between this particular oil from the Macondo well and the oil dispersant Corexit 9500A® is an urgent call to improve the knowledge we have on how to cope with oil spills and the role of oil dispersants in the process. The information of this posting was reproduced by more than 100 websites. Unfortunately our results were obtained at a time when the oil spill was already contained and the clean-up processes were done. Therefore, our results had no influence in those processes.

After publication of the scientific article in the journal "Environmental Pollution" online in November 2012, in print in 2013, we received various responses from the scientific community. Many of them were of praise, some of true constructive criticism, and a few of not-so-constructive criticism.

I became a scientist with the ideal of the search for truth having no economic or other impulse in twisting the results of the different scientific projects that I have implemented. I became a scientist for the joy of observing these little tiny microorganisms of enormous beauty that we called rotifers, their behavior, the way they respond to all changes in the environment, the need to protect that beauty and all the beauty that nature provides. If the results of our research, carried out with absolutely honesty, contribute to preserving rotifers and all other organisms in this universe, then I will happy with myself and my efforts.

References

Rico-Martínez R., Snell T.W., & T. Shearer. 2013. Synergistic toxicity of Macondo crude oil and dispersant Corexit 9500A® to the *Brachionus plicatilis* species complex (Rotifera). Environmental Pollution. Volumen 173: 5-10.

What the Narwhals Taught Me

Ari Daniel

*Ari Daniel has always been drawn to science and the natural world. As a graduate student, Ari trained gray seal pups (*Halichoerus grypus*) for his Master's degree in Animal Behavior at the University of St. Andrews, and helped tag wild Norwegian killer whales (*Orcinus orca*) for his Ph.D. in Biological Oceanography at MIT and the Woods Hole Oceanographic Institution. These days, as a science reporter for public radio and NOVA, Ari records a species he's better equipped to understand –* Homo sapiens. *Ari's radio stories have appeared on PRI's The World, NPR's Morning Edition and All Things Considered, Radiolab, Studio 360, Here and Now, Marketplace, and Living on Earth. Ari produces web videos and online science games for NOVA. He also helps run the Boston arm of Story Collider, a live storytelling show where scientists (among others) tell personal stories on stage. In the fifth grade, Ari won the "Most Contagious Smile" award.*

Back in the summer of 2004, I found myself north of the Arctic Circle on Baffin Island in Canada. Most of Baffin Island is wilderness except for a couple of towns. I was spending my time onshore with a science field team. We set up a handful of tents where we slept, ate our meals, and got our research equipment ready.

The first night I was there, I remember staring at a bucket of seawater that contained the cutting edge piece of equipment that I'd brought with me. I was looking at this bucket of water, hoping that it would work, knowing that it wouldn't, and wondering how I'd gotten myself into this predicament.

You see, I was studying whale communication and behavior, and I had been invited up to Baffin Island for three weeks after my first year of graduate school to do a project on narwhals. The most distinguishing characteristic of a narwhal is its tusk, which is a modified tooth that bores out and through the upper lip. (The myth of the unicorn comes from the narwhal since the tusk looks like a unicorn horn.) The lab that I was in worked with digital tags the size of cell phones that mounted non-invasively to the backs of whales, and they recorded the vocalizations and 3D movements of the animals. But you have to get the tag back to get the data. There's a release mechanism – a burn wire that allows the tag to float to the surface where you retrieve it with a dip net from a boat. The burn wire had only been shown to work in warm water. Never in cold water.

Before I left for Baffin Island, the project manager in our lab looked at me and said, "Don't f * it up." Working with these tags in the future depended on my successful execution of the narwhal project.

The idea for my field work was for me to put these tags on narwhals to see how they behaved in the wild. They tracked the three-dimensional movements of the tagged animal via onboard accelerometers and magnetometers, and they used hydrophones to record the acoustic environment (including the sounds produced by the animal). I was part of a team from Denmark, Canada, and the United States working on different research projects, and I was the only one from my institution who had been taught how to use and deploy the tags.

And so as I stared at the bucket of cold seawater on my first night on Baffin Island, I grew increasingly anxious as the burn wire refused to burn (which meant the tag would not release when it was on an actual narwhal). Since I didn't have an engineering background, I had no idea how to fix the problem.

The first couple of days, the weather was bad and we couldn't do any science. I was relieved – every day with bad weather was another day I didn't have to risk losing a tag.

During those few days of inclement weather, I got to know someone else who was in the field. His name was Mehdi. He was the lead engineer on the Crittercam at National Geographic and has also contributed a story to this book. He graciously offered to come up with a workaround solution – when the Crittercam released from the animal, it would pull the release mechanism of my tag, and both would come floating off. We had a plan.

The weather cleared on day 4 or 5. We were using a large net perpendicular to the shoreline to catch and then tag the narwhals before releasing them again. The first time we caught a narwhal, we brought it to shore to do the full workup. Its skin was beautifully mottled and its tusk was intricately textured. It felt like a gift to be that close to such a remarkable animal. Just before we were ready to release the animal back into the sea, we attached the Crittercam and my tag. And then we set the narwhal loose.

When the tag was at the surface, it transmitted a VHF radio signal. So I listened to my VHF receiver and I heard the characteristic beeping signal when the animal was at the surface, and silence when the animal was underwater. It meant the tag was successfully attached.

A couple hours later, it actually released. But by that point, unfortunately, a flurry of bad weather had kicked up and we weren't able to go and retrieve it.

The next day, Mehdi joined a couple of Inuit on their boat to look for our equipment, while I waited onshore. I hung out in the tent where we ate our meals until finally, a few hours later, Mehdi returned.

And he had nothing. My heart sank.

But then, he reached into his jacket and pulled out my tag. He was hiding it! I gave him a big hug. I skipped off to the science tent where I downloaded the data – a beautiful dive profile and a handful of narwhal vocalizations. I was thrilled. (Mehdi, however, was less thrilled since he had not been able to find his Crittercam – just my tag.) I realized that fieldwork is teamwork. Although I didn't have the expertise to fix my problem, Mehdi did and he was eager to help me.

Two days after we tagged that first animal, we caught our second narwhal. And this time, Mehdi decided we would tether the Crittercam and my tag together so we wouldn't lose one of them. We attached the tags and let the animal go. But after a couple of hours, our tags had not yet released.

3 hours passed.

Then 6 hours.

Then 10 hours.

There was still no sign that the tags were at the surface. Finally, a day or two later, we heard the constant VHF signal. Mehdi hopped into the Inuit boat and took off. And when he returned, he had both the Crittercam and my tag. I dashed to the science tent to offload the data, but it wasn't working. The tag wouldn't turn on. So I looked through the translucent housing containing the electronics, and the tag was packed with grit. It had sprung a small leak and the grit I was seeing was actually corroded circuitry. I used the satellite phone to call my lab back home, and they told me not to touch it. They would try to salvage the data and the tag upon my return.

I was sent to Baffin Island with three tags. One tag was down. I was apprehensive, and so over the next couple weeks, I declined to do more tagging, even as narwhal after narwhal was brought to shore, studied, outfitted with different instruments, and then released.

On our last night in the field, the forecast called for a tremendous snowstorm. My colleagues said it was up to me whether I wanted to put a tag on our last narwhal. I had been testing the release and it seemed to be working at least partially. I wanted more data so I took the risk, and attached the tag to the animal (without the Crittercam).

After we maneuvered the narwhal back to sea, the storm blew in with incredible fury.

The next morning, after such a harrowing night of weather, it felt like a new world. I fired up my VHF tracker and heard intermittent pinging – the telltale sign that the tag was still attached to the animal. But I had no more time left. The double propeller plane that was our only way out was inbound. I helped load the plane with all of our gear, and we flew to a nearby island base to wait for a few days. The tag wasn't necessarily lost, but we were moving farther away from it.

That first night, I couldn't even sleep. I awoke to a feeling of utter dread. Returning home with one tag clogged full of grit and another one lost at sea in the Arctic is pretty much the definition of "f *ing it up," to quote our project manager.

We posted a wanted sign with a reward for the lost Crittercam and tag, but the odds of our pieces of equipment washing ashore and someone stumbling across them were pretty much zero. At this base, there was a group of walrus researchers who were using a helicopter for their work. I called my advisor, and we dreamed up a plan where I would fly with them back to the scene of the crime, lean out the side of the helicopter, and reach down with a dip net to retrieve the tag. We realized it was too crazy of a scheme for me to try. But Mehdi said he would do it.

He was gone for half a day with the walrus researchers, looking for our tags. After hours of anticipation, Mehdi returned. And he had nothing. But this time, he really had nothing. While aboard the helicopter, he heard the signal of my tag getting stronger. He readied the dip net. He saw it, but it was still attached to the animal. It hadn't released. I had no choice but to return to my lab without that tag.

Back in the US, as I walked to the office of the lead engineer who invented and designed the tags, I imagined him telling me that I'd never lay another finger on one of these little machines. But instead, he said that I'd done the best job possible under those circumstances. And that they would work harder the next time to prepare whomever was heading out into the field.

And then, two months later, I received an email from one of the assistant engineers in the lab. He had taken that tag full of grit, opened it up, cleaned it off, and extracted 12 hours of data from the second narwhal. As a result, we were able to write a couple of research manuscripts on narwhal vocalizations and on how these animals turned upside down when they dove underwater. No one had described these behaviors before, and they offered hints into the role that their vocalizations might play and how they might hunt for food underwater.

My trip to Baffin Island allowed me to listen to and track narwhals in a new way. The Arctic doesn't yield its secrets easily, but the data I managed to retrieve offered a new sliver of insight into these animals.

It's been over a decade since I made those recordings, but that spirit of seek-and-discover remains with me. I'm a journalist, and now I aim to discover the beautiful stories about science nestled inside each of us – waiting to be told, ready to be held.

A Planetary Journey Through a World of Discoveries

Patrick Martin

Dr. Patrick Martin has been working since 2000 at the European Space Agency (ESA). He has over 20 years experience in the Space Sector field, notably in the positions of Research Associate; Deputy Project Scientist; Science Operations Management and Mission Management. He has been involved in a number of space projects and interplanetary missions, including NASA's Galileo mission to Jupiter, NASA's Near Earth Asteroid RendezVous mission to Asteroid 433Eros, and ESA's Mars Express, Venus Express and Rosetta missions. Patrick is currently Rosetta, Mars Express and Venus Express Mission Manager at the European Space Agency.

From innocent childhood to a seasoned explorer

Born in France in 1967 and being of a rather curious nature, for me it all started as a child with an insatiable, unexplored passion for science, construction and exploration in medicine, cars and trains, or astronautics, and through various media such as reading, television and computer games. Remember that this was at a pre-globalisation time when much of this media was not as developed and widespread as it is nowadays. It still was very much geared toward captivating an audience fascinated by discoveries in the late 70s of faster computers, new train, car and plane technologies, the Viking landings on Mars following the Apollo lunar successes about a decade earlier, and more.

This whetted my appetite for discovery, which then naturally translated into my school coursework, which followed a very diverse and meandering path, at times wavering between university and engineering degrees or between disciplines such as electronics, aeronautics, geology and astrophysics, albeit always keeping a sort of unconscious focus toward some sort of higher ground compared to very down-to-Earth activities. I say "wavering" because unlike some who seem to know right from childhood what they want to do with their life, I never had any preconceived ideas about what my future career would be about. All I knew was that at that time (in high school or early university) I would not be interested in what I would consider "conventional" jobs that this planet has to offer. Therefore I followed a convoluted path of trial and error which brought me from physics to telecommunications and from there to aerospace electronics, astrophysics and image processing of Martian landscapes.

It was from a small Mars data processing project in Toulouse, France, which turned out to be one of the best decisions I made, that I discovered a new world out there, dedicated to the study of our Solar System and in a broader way of the Universe we are in. I knew this was something that could both attract my interest and satisfy an untold thirst for exploration. By the end of my studies and by the completion of this project, I had learned to perform scientific data analysis and to research through massive amounts of existing corresponding literature. In other words, I had learned how to discover in a professional way, in addition to the natural curiosity for all aspects of life surrounding us as human beings.

It felt especially great at that time because I knew I could just get on a plane and go do research pretty much anywhere on the planet. Which is what I did; I moved from Toulouse to Hawaii for a postdoctoral Research Associate position. Many years later this has proven itself to be the path I needed to take. But make no mistake, I also strongly believe that without a number of opportunities or events that took place along this road to becoming a seasoned explorer, it may also have gone completely haywire and then who knows where I would now be or what I would be doing!

What was out there to discover?

I acquired through time a vivid interest in planetary science, in line with my rather unconscious attempts to somehow escape from human-centered life! This is in fact how I ended up reaching the Moon and Mars, our most well-known neighbours beyond our birth planet, via research. This discovery process started in a small planetary research laboratory in Toulouse, France, at the "Observatoire Midi-Pyrénées" (http://www.obs-mip.fr), with a pre-PhD project focusing on attempting to match ground-based, high-spatial resolution hyper-spectral data (http://tinyurl.com/hyperspectralimaging) onto the surface of the red planet. The concept of hyperspectral imaging, or imaging spectroscopy, is briefly described hereafter and in the Figure below: For each pixel in an image, a hyperspectral camera acquires the light intensity (radiance) for a large number (typically a few tens to several hundred) of contiguous spectral bands. See the figure 2 at http://www.cossa.csiro.au/hswww/Overview.htm. Every pixel in the image thus contains a continuous spectrum (in radiance or reflectance) which can be analysed in order to characterize the spatial units in the imaging scene in terms, for instance, of surface mineralogical composition.

This was new in the sense that this had been the first time an instrument designed purely for being used in the astronomical field of stars and galaxies had acquired planetary data from the Canada-France-Hawaii telescope located on top of the Mauna Kea volcano on Hawaii's Big Island. Right there was a very exciting and challenging project for me as a student, the outcome of which fulfilled some of the research goals set a few years earlier by Dr. Patrick Pinet who would later become my PhD advisor. This project quickly became a very positive first experience which reaffirmed my determination to persevere into this field of activities.

And that's how I embarked on a PhD, usually not an easy decision to make as this meant several years of intensive research work with an uncertain outcome and with the obligation to publish results in scientific journals. Even though it may seem easy to state this after the fact, the high interest of the PhD subject, "Optical and mineralogical heterogeneity of planetary surfaces through spectral study: the Tharsis region (Mars) and the Humorum impact basin (Moon)", convinced me to take the plunge into what then appeared to be the deep end for me. On top of it there was the prospect of the unknown at the end of the process.

This work, based on the theory of reflectance spectroscopy, comprised a first part dealing with the study of the optical, mineralogical, and thermo-physical properties of the Martian surface, in relation with the alteration processes and the proportion of exposed surface rocks, and a second part dealing with the optical, physico-chemical, and geological study of a lunar impact basin. The "Mars topic" consisted in pursuing the work that I had started a few months earlier (described above) but enhancing the preliminary results by looking into the surface composition of the areas on Mars resulting from the spatial matching of the telescopic hyper-spectral data onto the planet's surface. New near-infrared reflectance spectra of Mars had been derived from spectro-imaging measurements obtained at the Mauna Kea CFH observatory (Hawaii) during the 1990 Mars opposition, from a systematic survey of the Tharsis volcanic region of Mars. The primary goals of this investigation were to identify the physico-chemical origins of the observed spectral variations, to map the surface units based on their spectroscopic properties, and to draw inferences in terms of compositional heterogeneities of bright and dark materials (Mars' surface is very much characterised by bright and dark materials, depending on the surface alteration).

Detailed information regarding the ferric and ferrous mineralogy of soils was presented from my work, based on a reliable set of highly-resolved Mars spectra and the application of various spectral analyses to the dataset. Results of the modeling revealed that at the available spatial resolution, most of the investigated terrains could be explained by various combinations of clinopyroxene-dominated with hematite-dominated minerals. Detailed results from this paper are available through the following link: http://adsabs.harvard.edu/abs/1996P%26SS...44..859M. This discovery had the significance that we could infer specific mineralogy on Mars from low spatial resolution datasets from a ground-based instrument originally designed for astrophysical applications, which meant the overall goal of the experience was achieved.

The "Moon work" used multispectral data collected by the Clementine spacecraft which operated in lunar orbit in 1994. From this data I built a high-resolution image mosaic of the Mare Humorum region of the Moon and characterized and constrained the chemical composition of the region to establish a first estimation of both the mineralogy and maturity of the surface, on the basis of the spectroscopic properties.

Moreover, the combination of the results issued from the spectral mixture modeling with the geomorphologic and topographic information of the region led to infer some implications on the emplacement of the different types of lunar mare and highland units, on the contaminations of materials within these units, as well as on the stratigraphic variations of the crust related with the formation of the impact basin. This discovery contributed to further constrain our views and understanding of the evolution of the Moon.

The PhD results, albeit not ground-breaking, were satisfying and left a feeling of accomplishment before moving to the next stepping stone in my discovery process, postdoctoral research projects. This is when a setback occurred: I had been selected to do research work on the Mars-96 project in two different locations in the United States. Unfortunately, the Mars-96 spacecraft launch failed in November 1996, with the spacecraft ending its short-lived mission in the Pacific Ocean, taking with it my postdoctoral studies! But to the credit of Prof. Thomas B. McCord, with whom I had a Mars-related postdoctoral position lined up in Hawaii's Institute of Geophysics and Planetology (HIGP), I was rescued and went to work in Honolulu. Instead of working on a Mars mission, I did postdoctoral research on the analysis and interpretation of the NIMS (Near Infrared Mapping Spectrometer) data from NASA's Galileo mission to Jupiter. More specifically, I worked on analysing hyper-spectral data from the three icy Jovian satellites Ganymede, Callisto and Europa. This was a new world for me in many aspects: First the country of residence, not only the United States overall and academic culture, which for a French citizen represented both a change with a need for adaptation and at the same time a very exciting challenge, but also the multicultural world of the Hawaiian setting. Then obviously the language barrier played a role for a few months, which meant I benefitted from a free crash course in English from day one, and also learned to distinguish between the various accents of American English!

The HIGP institute itself was an interesting mix of people working in fields as diverse as volcanology (being near Hawaii's Big Island and volcanoes was certainly a bonus), lunar and Mars science, and outer planets studies with new results coming in from the Galileo spacecraft at the time I was arriving, and anticipating data to come some years later from the Cassini mission to the Saturnian system. During my time there in a subtropical paradise, quite a number of major discoveries were made from the results of the Galileo data analyses, one of the main ones being about finding evidence of hydrated salt minerals on the surface of Europa, closely matching magnesium sulfates and sodium carbonates, that may be reaching the surface from a water-rich layer underlying an ice crust. For details please see: http://www.sciencemag.org/content/280/5367/1242

This research and new way of life in a foreign environment turned out to be a very rewarding opportunity for me and allowed me to meet very interesting people, including my wife! Then one day I received an email from a Cornell University connection of my PhD advisor, asking me if I would be interested in working with him on an asteroid-bound mission. I saw the opportunity for a change and took it.

From Hawaii we moved to Ithaca, a small university town in New York State, where I started work as Research Associate at Cornell University and team member of the Near Earth Asteroid Rendezvous (NEAR) mission sent by NASA to the asteroid 433Eros in 1996 with the orbit insertion in February 2000. I have great memories of my time at Cornell, a place where astronomers have played major roles in NASA missions to explore the Solar System and distant universe, the late Carl Sagan being one of them. I could work there thanks to Dr. James F. Bell, collaborator and friend of my PhD advisor, with whom I enjoyed working very much on asteroid and Mars research. The NEAR spacecraft was the first to orbit an asteroid and answer fundamental questions about the nature and origin of near-Earth objects. The nearly two years spent there in Cornell led to quite a number of exciting discoveries, based on the analysis of the data from the Near Infrared Spectrometer on-board NEAR from the shape, structure and surface composition of the asteroid Eros (http://www.sciencemag.org/content/289/5487/2088), to the discovery of the East coast of the United States and Canada, including as well a few Mars observing trips back in Hawaii using NASA's IRTF telescope.

After finding out from my PhD advisor about a job opening back in Europe that would suit my profile, I applied for a job at ESA, the European Space Agency. Through a combination of experience, connections and serendipity, I moved to the Netherlands in 2000 to work on the Mars Express mission. As the first European Mars mission, Mars Express was launched in 2003 and has since greatly contributed to changing the way we see Mars in terms of composition, age and evolution of its terrains, providing a wealth of ground-breaking science datasets and results to the worldwide planetary science community. Mars Express highlights can be found here: http://en.wikipedia.org/wiki/Mars_Express.

From June 2013 I became Mission Manager for Mars Express and Venus Express, the latter of which recently completed its lifetime in Venus orbit after far exceeding its mission goals. Venus Express returned over 8 years of scientific data allowing many discoveries regarding the atmosphere and surface of our hot planetary neighbour Venus (http://en.wikipedia.org/wiki/Venus_Express). Through this link only a few of the discoveries are listed, and the best way to have a complete and detailed overview of the vast amount of Venus Express results will be to read the upcoming "Venus III" book whose publication is currently being planned. These results have a strong significance for the understanding of the structure and composition of the planet's atmosphere and its climate, as well as in terms of comparative planetology. This section described a "discovery" path I went through, reaching results through various opportunities that presented themselves, while at the same time keeping an eye on the "big picture" of why such achievements have a purpose, which is the scope of the next section.

Why discoveries are crucial for the human race to continue evolving

What now? Well, I just started as Mission Manager of ESA's Rosetta mission, a fantastic endeavour about 500 Million kilometres from Earth through flybys and orbiting of comet 67-P Churyumov-Gerasimenko. In November 2014 the mission caught the world's attention with the first ever comet landing. Rosetta is by itself in a different league in many ways: because of its impact on the public, because of its incredible technical and technological prowess, and because the scientific return and achievements, as ground-breaking as they already are, are potentially massive in terms of further understanding the origins of life on our planet. It is through the advancement of scientific knowledge and technological breakthroughs like Rosetta that the human race will pursue its evolution. There are in my view various degrees to which the discoveries we assist to or make have significance and benefits.

First, at what I would call the "local, immediate discovery level," we obtain results, sometimes ground-breaking, which have immediate consequences on our lives, whether these come from learning to monitor our planet's climate, from technological progress in areas such as telecommunications or transport, or simply from creating new ways of building infrastructure, cars or houses, all of which may have an impact on the future of our economy. Such advances become the most "concrete" applications for human activities as they can quickly be translated into new ways of living for instance via new technologies like internet-based devices.

Then I would see a "regional, distant discovery level," consisting in more fundamental findings but still having rather important consequences on our future as a human race. This takes place for instance through fundamental research, whether it is about curing cancer or confirming the relativity theory or demonstrating evidence of new molecules on a planet or small body in space via observation-based research aimed at constraining the scientific knowledge of the evolution of our Solar System.

Finally, as a sort of "global, abstract discovery level," in a way more difficult to grasp and maintain, I am a firm believer that a mind-set aimed toward discovery is the only and essential way for the human race to go on and be able to leave a planet which at some point may not be as harbouring and hospitable as it has been in the past. Many threats are upon us, with most of them being ignored or not fully and seriously acknowledged, from overpopulation to politically motivated conflicts to biological threats and to external threats. This is why education, reading and the internet are crucial means to be maintained and expanded so as to encourage such mind-sets toward perpetuating discoveries.

This article has reviewed a few of the discoveries that I went through in both my working life but also as a human being walking the blue planet. Each scientist or engineer has their own path through the world of discoveries and has often many possible ways of describing them or even living them. The important thing is not how we describe or live them, but rather to keep the pace of discoveries. Let's hope our planet and the species living on it will live long enough to be able to make this happen, and then go on to discover other worlds!

We hope you enjoyed the ride and learned from these compelling stories.

Read Marc Guttman's other books - www.Why-Peace.com & www.WhyLiberty.com

www.ingramcontent.com/pod-product-compliance
Lightning Source LLC
Chambersburg PA
CBHW020153200326
41521CB00006B/346
9780984980239